EPIDEMICS LAID LOW

EPIDEMICS LAID LOW

A HISTORY OF WHAT HAPPENED IN RICH COUNTRIES

Patrice Bourdelais

Translated by Bart K. Holland

THE JOHNS HOPKINS UNIVERSITY PRESS I BALTIMORE

Originally published as *Les épidémies terrassées: une histoire de pays riches* © 2003 Éditions de La Martinière (Paris, France)

© 2006 The Johns Hopkins University Press
All rights reserved. Published 2006
Printed in the United States of America on acid-free paper
9 8 7 6 5 4 3 2 1

The Johns Hopkins University Press
2715 North Charles Street
Baltimore, Maryland 21218-4363
www.press.jhu.edu

Library of Congress Cataloging-in-Publication Data
Bourdelais, Patrice.
    [Épidémies terrassées, English]
    Epidemics laid low : a history of what happened in rich countries /
Patrice Bourdelais ; translated by Bart K. Holland.
        p. cm.
    Includes bibliographical references and index.
    ISBN 0-8018-8294-X (hardcover : alk. paper)—ISBN 0-8018-8295-8
(pbk. : alk. paper)
        1. Epidemics—Europe—History. 2. Medical policy—Europe—
History. I. Title.
    RA650.6.A1B6813 2006
    614.4—dc22   2005018235

A catalog record for this book is available from the British Library.

CONTENTS

INTRODUCTION TO THE ENGLISH-LANGUAGE EDITION

Since this book first appeared in France in 2003, the world has seen at least two major epidemic alerts. The first, an outbreak of SARS (severe acute respiratory syndrome), occurred in late 2002 and early 2003, disappearing by the end of the year. The contagion was first recognized in Canton and spread some months later to Hong Kong, where the main airport served as a conduit to the rest of the world. SARS created immense fear, significant mobilization of the authorities—even if they did react more slowly than the citizenry—and an economic loss of several millions of dollars in Asian countries. As for the number of victims, it is difficult to give a global estimate because of the probable underreporting of the deaths in China. The World Health Organization (WHO) has suggested that SARS was responsible for 774 to 916 deaths, mainly in China, Taiwan, Southeast Asia, and Canada. (Toronto had several hundred cases.)

The second worldwide alarm, again occurring in Southeast Asia, took place in 2004, when the chicken flu caused "only" thirty deaths. This outbreak worried the health authorities because of the possibility of a direct contamination from chicken to people without the traditional intermediate stage in pigs, which usually allows some months for the preparation of new vaccines each year.

Asia seems to be the source of numerous new epidemics. The subtropical climate is a factor, as is the rise in population, which has grown by 40 to 67 percent during the last twenty-five years. People and farm animals are living in ever-greater proximity. These three main factors probably contribute to the appearance of new strains of influenza. The biological nature of the virus is also important; each year it modifies its structure.

A small change in temperature can also modify the epidemiological balance. In 1993, a new epidemic appeared in the Southwest United States (New Mexico, Arizona, Colorado, and Utah); several people died. The cause was the hantavirus, which is spread by the deer mouse. The proliferation of that mouse species was the direct consequence of the El Niño phenomenon, which produced a series of warmer years.

Global warming has already had an effect on some infectious diseases and epidemics. Major climate changes did occur in the past, but the ways of producing and consuming in rich countries are partially to blame for the present situation. Yellow fever, malaria, and dengue can now enlarge their territory because a one- or two-degree increase in average temperature can have a significant influence on mosquito activity. As a direct consequence of warming temperatures, malaria and dengue are now present in the mountains of Africa and in Central America. For the same reason the West Nile virus reached New York City in 1999. Viruses and pathogenic germs can make their own way without any help from human activity. Most frequently, however, epidemics are the result of interactions between ecological changes that favor germs and human activities that facilitate the reproduction of mosquitoes, for instance (swimming pools, sewers, puddles), or the circulation of germs (as with legionella in air-conditioned buildings). Other human activities, such as long-distance commerce and air traffic, also have an effect.

Considering these epidemiological events and worldwide mobility, it seems that our world, even in the wealthiest regions, is not free from the ancient threat of epidemics. This situation is a big disappointment after several centuries of actions to control and eradicate contagious and infectious diseases. Disappointment may lead to an exaggeration: the worldwide perspective can even give the (false) feeling that epidemics are more numerous and dangerous today than they were some decades ago.

In the public health field, there is widespread concern about increasingly rapid exchanges between all regions of the world, even the most isolated. Moreover, historians know quite well that each step toward the globalization of trade and the economy has led to a new generation of epidemics (such as plague, syphilis, yellow fever, cholera, and influenza). One might expect that a wave of new epidemics would appear now, because of the new scale and intensity of our globalization. These two recent alerts have clearly led the public to realize that globalization and the density of air traffic may have some very bad consequences for health security, even in rich countries. On the other hand, people in Southeast Asia and China have realized that their economic prosperity, which is linked to trade with many countries, is dependent on their public health. For the first time China has had to accept the presence of WHO experts. Epidemics have always been an incentive for institutional intervention: to obtain information that would not otherwise be disclosed, to monitor behavior, and to control the implementation of policies. This was the case in China during the SARS epidemic. It is impossible to participate in international trade and remain deaf to the requirements of an international institution such as the WHO.

These two alerts (SARS and hantavirus) occurred in a specific historical context. Since the early 1980s, the world has had to cope with an incredible new epidemic, AIDS (acquired immune deficiency syndrome). The United States and Western Europe were the first to experience thousands of deaths and ill people each year. In less than fifteen years, however, the number of infected people and the number of deaths decreased because of the implementation of preventive measures and the development of new medical treatments that made AIDS a chronic rather than an acute disease. In France, a peak was reached in 1994, with 4,200 new cases and 5,700 deaths; in recent years the numbers have fallen to 1,400 cases and 500 deaths. All of the rich countries have been able to control the epidemic in less than fifteen years. Yet AIDS still leads to demographic and social catastrophes in less wealthy countries, and on a completely different scale than that experienced in rich countries. In South Africa, the proportion of infected persons is estimated to be between 18.5 and 25 percent of the entire population (between 4.5 and 6.2 million persons). The number of orphans due to AIDS is probably around 1.5 million. Many other African countries also have very high proportions of infected persons, as does Thailand. China's data remain obscure, but it seems that the number of infected people is higher than declared (probably in the millions). If older epidemics (such as tuberculosis, measles, and malaria) are also considered, it is clear that the improvement observed in the Western world has not spread elsewhere.

This book is focused on how rich countries effectively controlled epidemics of infectious diseases. For a historian, the answer to this question is obviously more complicated than the economic level or medical services available in the various countries. Without ignoring these factors, this book has a different perspective. It emphasizes all the ways in which, and the reasons why, our societies have fought against epidemics without being able to eradicate them. We know today that, because of the relationships between germs and human beings, total victory will forever elude us. Yet we can achieve an extraordinary decrease in mortality from infectious disease.

The factors influencing epidemics and their control are so numerous and so closely linked that it seemed logical to look back in time to see how European societies confronted the Great Plague in the mid-fourteenth century. It is clear that the authorities at that time pursued many coordinated actions to control the plague. Ever since this "Black Death," one of the constants of the elites (political, social, medical) and of the population at large has been to try to fight against epidemics by all means, and in particular by creating specific institutions, which are not necessarily concordant with our contemporary scientific knowledge but are ultimately efficient. Quarantines, sanitary cordons, lazarettos, even if they were necessarily porous, did change the scale of epidemics (and were used recently in the SARS outbreak). The determination of governments and their measures to protect cities, provinces, and countries from danger developed over five centuries, beginning in the fourteenth century.

Economic, political, and social trends converged at the beginning of the nineteenth century with new medical theories leading to the abandonment of the old regulations against plagues. Faced with cholera epidemics, governments turned to new behaviors more compatible with the new rules of commerce and with the new demands for democratic reforms. Efforts were then focused internally, within the European countries, on urban districts with poor people; new "borders" were found *within* society. The working classes had to be acculturated to new medical knowledge and had to adjust their behavior according to it. At the end of the nineteenth century, to be ill often meant to be held responsible for disease because one's behavior was considered to be the cause (especially for tuberculosis and venereal diseases). At that time, public health campaigns and sanitary institutions focused on domestic reforms. Western European governments tried to impose precise controls on people immigrating from Egypt and Turkey, and the United States asked Western European countries to control their ports and not send ill people overseas. Ellis Island is well known all over the world for the large-scale "traditional" sanitary control that took place there.

It was probably the Spanish flu that severed the connection between illness and personal responsibility. In the second decade of the twentieth century, between 25 and 45 million deaths occurred around the world within less than two years. The epidemic was so powerful and widespread that it was obvious that the factors influencing its spread went beyond individual behavior. Microbiologists were looking for germs, but the small size of the virus prevented identification of a flu organism until the 1930s. The Spanish flu is still an active subject for research because scholars would like to know the precise structure of the virus. It might then be possible to understand why it killed so many people. Scientists still consider it a potential danger, highly adapted to our world.

Even if it was an illusion to believe in the eradication of epidemics, it is also clear that this conviction, this ideal, has been one of the keys to the success of control efforts. Today, in spite of the deceptions or fears generated by new epidemics—and there will always be new ones—it is possible to analyze why the control of epidemics is not complete. But our main goal should be to improve the situation in poor countries. The majority of human beings are still not safe from epidemics that seem to belong to the past in rich countries. The new frontier is there. And the historical experiment of the rich countries provides evidence about the most efficient measures for controlling epidemics—we need not only vaccination and antibiotics but also better food and housing, better personal hygiene, and better organization of public health by elites for both humanitarian and economic reasons. We also know that the new immigrants in large cities have to be integrated, in social and in economic terms, and have to have a job that supports their families. It seems clear that below a certain security and wealth threshold, a person is unable to think of the future and incorporate attitudes that can prevent illnesses.

The historical experience also provides evidence that local officials are vitally important in achieving reform by implementing new institutions for care or social relief and acculturating people to new rules of hygiene and prevention. Successful reform requires these institutions and some decades of public education. Although the solutions to these problems seem clear, the many contradictions and tensions in society interfere with their application. Political involvement is crucial, which is probably why thousands of volunteers from the richer countries are working with nongovernmental organizations to implement medical and social relief, even in seemingly hopeless situations. In these efforts we see the best of human nature. Working as a volunteer in a poor country is an example of human generosity. In this book on the history of fighting against epidemics and infectious diseases, this aspect of human history plays an important part.

From a historical perspective, it is clear that the battle for improved control of infectious diseases is never won. Examples in northern countries (the return of epidemics in Russia and increased mortality in central Europe after 1989) and in countries of the south (high epidemic mortality because of wars, administrative disorganization, poverty, and sexual tourism) show the fragility of the progress made in fighting epidemics. Most of the world's people live in countries where conditions are not ideal and the economic resources necessary to fight infectious diseases are lacking. The fact that they constitute a major epidemiological risk for rich countries can be a source of hope. Moreover, the historical experience of wealthy nations shows that gaining the upper hand over epidemics is possible, although it requires a daily struggle in the face of social disorganization.

PATRICE BOURDELAIS

EPIDEMICS LAID LOW

Legionella, salmonella, and listeria are the three infectious organisms most commonly mentioned in newspapers in the Western world, yet their names were totally unknown to readers just a few years ago. New epidemiological threats are described in our daily papers in a manner out of proportion to their statistical rarity and the small number of victims they affect. This is to be expected because the press is on the lookout for unusual, sensational events and especially for news that the reader is concerned (or, better yet, worried) about. The discovery of each new disease calls to mind periods in history when the inability to control epidemics led to terrifying death tolls; insecurity sets in. Our achievements of more than a century—in bacteriology, in vaccination, and in methods for safeguarding the public health—aren't they bound to protect our societies from such dangers? Aren't the dangers from epidemics all the more worrisome when they appear in the very places where technical progress has been spectacular in the last quarter century (such as industrial food production and the supply of hot water and air conditioning)? What's more, the new diseases affect aspects of our society that are highly valued, namely consumption, comfort, and medical care, especially hospital-based medical care. These diseases make us remember the old idea that all progress contains the potential for the opposite. But they also make us remember that we need to assess the risks involved, so we can come up with adequate defenses. The new potential for epidemics creates the fear of a return to a past that we had hoped was gone forever—or the fear of arriving at a situation like that in regions of the world that are not so well off as we are.

In fact, in poor countries, most of the great epidemics of our Western past, except for smallpox, are still around, carrying off a significant proportion of children and making the lives of adults uncertain. So many factors kill the hopes of those who aspire to the measured, rational attitudes of our Western societies, where we pay so much attention to the care and maintenance of a healthy body. So much stands in the way of even the aspiration to life itself! One of the limits to solidarity with developing countries stems from this disparity. But would our fellow citizens be willing to give up the annual improvement of their own life expectancies so that significant resources could be transferred and used to help impoverished populations?

AIDS (acquired immune deficiency syndrome) was, for more than a decade, a principal focus of the press and of government authorities, yet now it seems to have slipped into the background, although no vaccine or definitive treatment is available today. Why is AIDS being ignored while legionella bacteria living in the cooling towers of your neighboring apartment building and listeria in the food supply seem to be the talk of the town? Probably because, once the vector and the mode of transmission are known, it seems that an epidemic must be under control, even if it continues to claim many victims each year. When it comes right down to it, the subconscious age-old distinction remains between "innocent victims" (in the case of AIDS, acquired from a blood transfusion or the use of blood products) and those responsible for their own infection (through drug use and homosexuality). Because the modes of transmission are well understood today and the blood supply has been made safe, people with AIDS are clearly less free from stigma and blame than they were fifteen years ago. This factor accounts for the current indifference of the public and the silence of the press. In public opinion, hasn't the epidemic already been stamped out? Isn't it already a thing of the past?

This perception is all the stronger because one of the definitions of a rich country is one that has succeeded at conquering epidemics one after the other and that has markedly reduced overall mortality. Moreover, in tables of data showing the circumstances of various countries, you find infant mortality rates and life expectancies right next to average annual income. These figures are presented to indicate the rank order of countries with respect to their level of development. Such tables also reflect choices because there are other variables, such as income disparities or gender equality, that are excluded. The implicit, powerful assumption is that development requires the suppression of major epidemics. When epidemics strike poor countries today, they instantly evoke images of the past in the collective imagination. The most familiar images involve diseases against which vaccinations and antibiotics have had spectacular success. These diseases include tuberculosis, whose ravages were widespread until the 1950s; long-feared ailments like diphtheria, typhoid, and polio; and smallpox, now eradicated from the surface of the earth. But the great epidemics of the past—plague and cholera—have also shaped beliefs and behaviors over the centuries. Experience with these two ailments formed the basis for the two principal methods for battling epidemics: protection against incursions of disease from elsewhere and amelioration of urban conditions that are conducive to disease. First came lazarettos [isolation hospitals],* quarantines, and military blockades, followed by a wave of efforts aimed at improving sanitary facilities and modifying health-related behaviors, at a time when people were demanding more health-related interventions.

This book offers a history of the dynamics that link populations, doctors, and sanitary policies in the struggle against epidemics, but the focus is not on an exhaustive retelling of the medical and epidemiological details. The goal was not to establish the demographic toll of epidemics or to delineate the evolution of the social role of the physician, nor was the intent to recount the history of the establishment of governmental social welfare institutions. Instead, elements of all these topics are discussed as needed. For example, it is impossible to understand fully the influence of antibiotics in the struggle against major epidemics without taking into account governmental social welfare institutions, improved standards of living, and the rise of the pharmaceutical industry.

The emphasis on the politics of the struggle against epidemics has certain consequences for the chronology adopted in this book.

---

*Words appearing within square brackets are translator's additions.

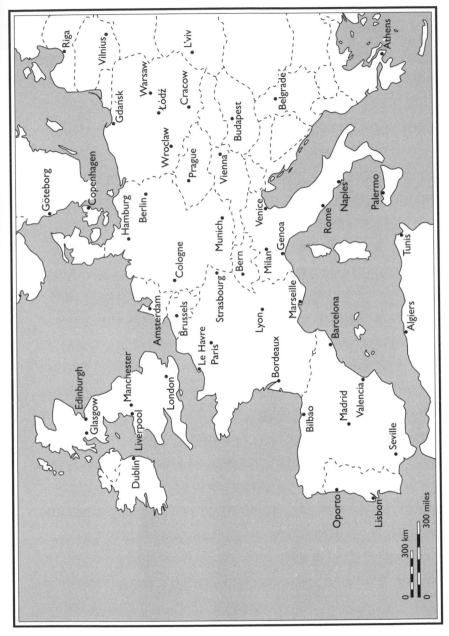

Map of Europe

We start with the bubonic plague of the Middle Ages, which engulfed Europe in the mid-1300s, because the plague gave rise to the initial concept of the epidemic in the mind of Western people and because public policies to protect cities originated at that time. We deal with the last outbreak of smallpox, in southern France in 1720–22, and our chronology continues with the origins of the policy of voluntary smallpox vaccination. The policy is seen as an adaptation emerging from the concerted efforts against the disease, made possible by the rise of royal governments in the seventeenth century, by new health practices, and by a newly emerging medical establishment.

The next development was marked by the abandonment of traditional plague protection policies at the time of the first European cholera epidemic (1831–32). New policies arose, which were to change yet again with the advent of the bacteriological revolution, the development of various vaccines, and, most recently, the widespread use of antibiotics.

Another consequence of the approach taken in this book is that the story ranges over a wide geographic area and naturally extends beyond France. In the medieval period, Italian cities instituted the first coordinated measures for the control of plague. In the nineteenth century, during the shift from one protective system to the next, the Russian experience and that of the German states—as well as decisions made in England—were crucial. In Sweden, new solutions were developed to advance smallpox vaccination and the battle against syphilis. Then, with Western European powers instituting a policy of checking and screening on the shores of the eastern Mediterranean, we need to take a detour to Constantinople, Mecca, and Alexandria.

In focusing on the reactions of populations and the epidemic control efforts resulting in the present-day situation in developed countries, this book argues that it is most important to understand the multiplicity of factors that produced this progress. These factors include the overall policies of national governments, urban programs, the demands of the people, the role of physicians, scientific discoveries, and access to care. This list includes factors unavailable to those seeking to reduce mortality in the poorer countries today.

1

In the modern imagination, the plague remains the foremost example of the scourge of epidemic disease. In 1933, Antonin Artaud wrote of plague's currents of unpredictable, frenzied energy.

> The last ones alive went crazy. The usually respectful, dutiful son killed his father, while the sexually repressed sodomized friends and relatives. Lovers of luxury became ascetics. Misers threw their gold out the windows by the fistful. The war hero burned down the town he had previously protected at great cost to himself. The dandy got all dressed up and went for a stroll on mass graves . . . How can we understand the erotic motivations among those who recovered and did not flee, seeking instead to grab reprehensible sexual pleasures from the dying or even the dead, half-crushed under the pile of dead bodies where chance had deposited them.[1]

The nightmare of plague, whose danger has been feared for centuries, called attention to the fragility of life; it increased religiosity, shaped disease control measures, and influenced literature and art until recently. Central to this long-lasting impact were the Black Death of the fourteenth century and successive waves of plague over the centuries. Plague was far from the first epidemic to affect humankind—after all, contagious diseases had afflicted the human species since its very origins—but it was the first episode in a long story of misery that was to last until the eighteenth century.

From the Plague of the Philistines to Justinian's Plague

Striking outbreaks of sudden, frightening, deadly diseases—often associated with wars—have been described in writing for millennia. Terrible epidemics are mentioned in sacred texts of the ancients and, later, in historians' chronicles. The first book of Samuel in the Old

Testament provides the earliest detailed description. The text describes the calamity that befell the Philistines when they seized the Ark of the Covenant after defeating the Israelites at Ashdod in 1141 BCE. Although all retrospective diagnoses are speculative, the disaster was probably an outbreak of severe dysentery. The second book of Kings tells the story of another huge epidemic during the siege of Jerusalem by the Assyrian king Sennacherib in 701 BCE, when his troops suffered heavy losses, most likely due to a virulent form of malaria, and could not rally to attack.

One of the best-known epidemics of ancient times was the "plague of Athens," which occurred in 430 BCE. The story of this outbreak became famous for several reasons: Thucydides described it, Diodorus of Sicily provided casualty figures, and Pericles died in it. Although called a plague, the disease was probably typhus, and this incident, too, is linked to war.[2] The outbreak of disease occurred among the Spartans while they were pursuing the conquest of Athens; the mortality rate was so high that they had to abandon that goal until the illness passed. Similarly, Diodorus of Sicily left us a precise description of the illness that decimated the Carthaginian army during the siege of Syracuse in 396 BCE. The "plague of Syracuse" was probably either typhus or dysentery, the scourges of every army campaign from late antiquity to the twentieth century. The Antonine plague (AD 167–180) was probably brought to the western Mediterranean by the victorious armies of Lucius Verus and spread by large public celebrations in Rome. It ravaged the Italian peninsula and Gaul for fifteen years, making Marcus Aurelius's battles against the Germans more difficult and finally killing him in 180. The physician Galen followed Hippocrates' precept on epidemics, which advised fleeing "cito, longe, tarde" (immediately, to a great distance, for a long time). Although Galen visited Rome in person from time to time during this period, Galen's description of this disease is so imprecise that it has been interpreted variously to indicate smallpox, scarlet fever, and dysentery.

"Around 542 an epidemic broke out that practically eradicated the entire human species," wrote Procopius, the Byzantine historian of Justinian's reign.

It started at Pelouse, in Egypt, and spread from there in one direction toward Alexandria and the rest of Egypt and in another direction toward Palestine. It began at seacoasts and made its way progressively inland . . . There was no effective means of stopping it in time, nor of preventing it from running its inevitable course . . . Those in whom the buboes increase in size the most, and develop pus, tend to survive . . . The epidemic at Constantinople lasted four months; the death toll grew to five thousand a day, finally reaching ten thousand and even more.

This epidemic—"the plague of Justinian"—struck the West and the Middle East several more times until the eighth century. Gregory of Tours, in his *History of the Franks*, mentions its occurrence in Arles in 549, in Clermont in 567, and in Lyons, Bourges, Chalon, Dijon, and many more places. This epidemic was sufficiently well described that it can be accepted as the first documented outbreak of bubonic plague. There is frequent mention of inguinal and axillary buboes. Gregory of Tours is explicit: "A lesion starts in the groin or armpit and is like a snake whose venom affects the patient so strongly that, by the second or third day thereafter, the patient breathes his last, completely out of his senses."[3] Around 570, a fierce epidemic of smallpox, possibly the first in Europe, added its effects to a third wave of the plague.

For two centuries, the countries bordering the Mediterranean saw their populations decimated time and time again by these deadly waves, demographic consequences that were striking even to contemporaries. On the other hand, it seemed that northern Gaul, the Germanic lands, and the British Isles were spared, although a "yellow plague" struck the lands across the Channel several times from the mid-sixth to the mid-eighth centuries. This variability even suggested new explanations for the ebb and flow of political power: "It's an overstatement, no doubt, but one might argue that the plague of Justinian, after providing perhaps a partial explanation for Mohammed, could also explain the rise of Charlemagne."[4] After all, a demographic depletion was caused by the epidemic, and the corresponding collapse of tax revenues in the Byzantine Empire may have helped the Arabs succeed in North Africa. Also, since northern Eu-

The spread of the Black Death across Europe in the fourteenth century.

From *The Black Death: A Biological Reappraisal,* by Graham Twigg (London: B. T. Batsford, 1984). Reproduced by permission.

rope was spared by the outbreak, its power could be used to the disadvantage of southern Europe and the Mediterranean.

After this period, the documentation is sufficiently rich to make it possible to assess how these terrible waves of mortality affected the attitudes and beliefs of the population. People probably became more obedient to Christian beliefs and practices, and the feeling of waiting for Judgment Day became rooted in their minds; they also saw original sin as an explanation of the disasters that, until then, had been seen as a more general expression of divine anger. They

adopted the image of an angry God and took up behaviors consistent
with apocalyptic and millenarian attitudes. During the winter of 589,
the plague of Justinian hit Rome hard. When Pope Pelagius II died of
plague on February 8, 590, the Romans were terrified. His successor,
the famous Saint Gregory, organized litanies and, in April, held a
procession during which the Angel of Death appeared at Hadrian's
villa, complete with bloody sword. The plague stopped immediately,
and thereafter the villa was called the House of the Holy Angel. More
generally, for the church authorities these plagues offered an opportunity to carve out new symbolic spaces in the cities—to claim pre-
eminence for Christian sites by organizing pilgrimages and processions, such as the one to the tomb of Saint-Julien de Brioude.[5] Saint
Gall introduced new prayers, the Rogations. All these manifestations
of collective contrition were intended to strengthen the social cohe-
sion that had been sundered by the pseudo-prophets and rabble-
rousers who preyed upon the fearful and led revolts against the rich
and powerful. The objective was also to avoid spiritual pollution and
to promote the idea of Christian purity. During the eighth century,
the plague of Justinian vanished, without leaving behind any permanent reservoir of recurrent infection.

The Black Death

After an apparent absence of six centuries, the Black Death returned
in the Middle Ages. All the chroniclers believed that they were faced
with a new disease, which warranted their detailed descriptions.
That's one of the reasons that some historians, most recently Samuel
Cohn, have cast doubt on the accepted wisdom that the Black Death
was really bubonic and then pneumonic plague.[6] The Black Death
was so different compared to what had been seen before that the
plague of Justinian was not used as a model to understand, protect
against, or fight the epidemic. Quite the opposite: the Great Plague
of 1347–48 provoked the development of several new models, which
were to endure for several centuries. These included both plague (as
one of the main symbols of collective misfortune) and distrust of the
East (seen as the source of all pestilence). At the same time, plague
provoked the development of the key measures used to control epi-

demics, measures that were instituted in the second half of the four-teenth century. These were maintained, for the most part, until chol-era struck northwestern Europe in 1831–32 during the course of a pandemic (an epidemic affecting several regions). Writings and im-ages from the period show the striking effects on people's minds of frequent, massive, deadly outbreaks. Of course, the Latin word *pestis,* which was widely used until the seventeenth century, was used to indicate any of the great epidemic diseases, such as plague, typhus, smallpox, or dysentery. But bubonic or pneumonic plague weakened populations that they did not destroy, thus increasing the mortality caused by more endemic diseases. Plague served as a reminder of the fragility of life. For several centuries, plague was to become the central epidemiological regulator: plague's cyclical recurrences in-fluenced demographic development in the European population and also had an influence on how people thought about life.

The Price of Growth

Late in 1347, a dozen Genoese ships escaped the Mongols' siege of the city of Caffa, in the Crimea. Wherever they came into port, they left behind a terrifying souvenir, particularly in Constantinople and, a little later, in Messina. In these cities, devastating epidemics struck as soon as the ships tied up. Three or four of the ships, turned away at Livorno and Genoa, were finally allowed to dock at Marseilles on All Saints' Day. Marseilles scorned the rumors that the ships were carry-ing a terrible illness. The intense trade war with the Italian markets led Marseilles to disregard the risk—and we know the consequences. Plague entered Marseilles and brought so many deaths that the sur-viving inhabitants fled, contributing to the spread of the disease, which, over the winter, began to appear in the pneumonic form. The bacterium no longer took shelter in fleas, spreading instead directly from person to person by coughing. The epidemic now spread faster, and mortality reached a new high wherever it struck. For example, one street in Marseilles lost almost all its inhabitants within a few weeks. Between late 1347 and the first half of 1348, the epidemic was established throughout Languedoc, moved toward Bordeaux on one side and Lyon on the other, and spread in Provence. Finally, the main cities of the Italian peninsula were hit: Genoa, Pisa, Venice, Lucca, Florence, Bologna, Sienna, Padua, Orvieto, Naples, Verona, and Parma all suffered.[7] Although not the first epidemic of plague to extend to Western Europe, this plague affected so many places that it became a true pandemic. It was much more serious than the plague of Justinian that had affected the early Middle Ages.

The expansion and growth of trade between Europe and the nations on the eastern shores of the Mediterranean provided the basis for the westward advance of this new epidemic. Above and beyond the twelve Genoese ships that were so perfectly situated to start an outbreak, the westward push of the Mongols created a broader flow of traffic from central Asian regions, where plague was endemic, to the shores of the Caspian Sea. From both northern and southern shores, this traffic continued farther westward. Most of the Mediterranean islands were infested with plague by 1347. Moreover, rapid population growth in many areas in Western Europe and the growth of cities meant that grain had to be imported from the eastern Mediterranean to feed the populations concentrated in urban areas. As the population grew and lived more closely together, trade also grew, as did the imbalance between population and food supplies. Thus, many factors were favorable to the spread of microorganisms throughout the Mediterranean region and the consequent spread of epidemic disease through all of Western Europe.

In four years, the plague traveled to every country in Europe, including Russia. Here was a true human catastrophe, causing an unprecedented demographic loss. Emmanuel Le Roy Ladurie has compared it to the effect of the atomic bomb on Hiroshima, with its sudden, extreme mortality rate and lasting consequences.[8] Historians have not been able to calculate exact mortality figures for the medieval plague, given the limited quantitative documentation for the more distant past. Nevertheless, all available indications (such as increasing numbers of wills and death records) point to exceptionally high mortality, as reflected by changes in overall population counts as well. Most Italian cities lost close to a third of their populations. French cities struck by the plague lost between a third and a half. For example, Albi had 2,669 "heads of households" in 1343 and 1,200 in

1357; in Provence, Aix lost 676 of 1,486 "heads of households" in the ten years between 1345 and 1355. Plague decimated the ranks of city officials but attacked members of religious communities even more harshly: all 150 brothers in a Marseilles religious order, the Cordeliers, died, as did all Cordelier brothers in Carcassonne; in Montpellier 7 brothers survived of 140. Clergy not living in religious orders were relatively less affected, even though losses were around 50 percent, according to data available from various locations in southern France. The demographic effects of this first epidemic were reinforced by successive waves, which occurred on average every nine to eleven years until 1534. After that, they occurred every seven to thirty-one years until 1683. In the mid-fifteenth century, at the close of the Hundred Years' War and thanks to the movements of troops, which helped spread all epidemics, the French population was smaller than it had been since the year 1000.

For four centuries, population growth in France was at a standstill because of the interacting effects of various diseases, plague foremost among them. Plague, however, was not alone in cutting lives short. Over the course of several centuries, an equilibrium among diseases became established, in which plague was central; periodic waves of that disease certainly limited the expansion of other illnesses. At times in epidemiological history, a newly emerging ailment like tuberculosis competed with another, more established disease, like leprosy. Sometimes a new scourge became endemic—constant but less prevalent and virulent—and a new equilibrium was attained, allowing even newer illnesses to occupy center stage.[9] The predominant illnesses at various points included typhus, smallpox, syphilis, and even the English sweating sickness. The period during which new diseases were emerging and superseding old ones came to a close toward the end of the seventeenth century, giving way to a long transitional phase during which epidemics gradually became attenuated over all. This was followed in turn by the emergence of a "modern" pattern of mortality.

The Black Death not only caused demographic stagnation; it also disrupted economic and social life for a very long time. City dwellers took flight, and mortality was high among magistrates and notaries, physicians, and those involved with food supply and preparation (because food drew rats). These factors had a huge negative influence on daily life and transformed poorer workers into beggars. Once the epidemic was over, those whose income derived from shares of trade and agricultural production (such as princes, lords, prelates, and religious orders) suffered an immense loss of resources that threatened all the activities they financed. A huge number of new people, often young and poorly trained, assumed positions as bishops and in lay and religious orders. The great plague disrupted government, the economy, and social and cultural life; on the other hand, it made way for a restructuring of local governments.

Decisions to Protect Health

Faced with this huge challenge, the authorities took action, particularly in the Italian cities. In March 1348, the Grand Council of Venice decided to elect a committee of three intelligent men from among its members and charge them with responsibility for "making all decisions to protect health." The following month, Florence, too, chose officials specifically charged with the same task, making rules that call to mind the *statuti sanitari* promulgated for similar purposes in 1321–24.[10] The key measures that were instituted focused on regulating the markets, on verifying the origins of merchandise, and on preventing the resale of clothing that had belonged to victims who had died of the plague. In Pistoia, in June, the first special corps of gravediggers was founded. In Orvieto, in August, one of the plague inspectors was a doctor who was receiving four times the salary that had been in effect the previous year.[11]

Information on the progress of the epidemic began to be gathered in a systematic way. The Duchy of Milan created a network of checkpoints along thoroughfares and at city gates and did not hesitate to completely isolate any houses with plague victims inside. Nor did the authorities hesitate to demolish those houses, burying the dead in the process. In 1374, Bernabo Visconti ordered all those suspected of harboring plague infection to be forbidden entry to the city until they had been isolated for ten days outside the city walls. He also required that those caring for the sick report the names of those who had plague. Toward the end of the fifteenth century, a sort of health passport—called a certificate of individual health—was in use everywhere in northern Italy. The existence of this type of certificate is also documented for Provence, at Brignoles, but only after 1494. Such certificates were issued by each locality visited and showed the final destination of the traveler as well as his or her age and hair color, allowing health officials to determine whether the new arrival had crossed areas contaminated by the plague.

In 1374, Genoa and Venice refused to let into port ships that had come from infected areas. In 1377, the town of Ragusa required a month's isolation before ships could enter the port. Soon thereafter, in Venice, this was extended to forty days, on the basis of the Hippocratic doctrine that considered the fortieth day as the last possible day of illness for acute diseases such as the plague. Thus, quarantine—the protective strategy that was to be used until the middle of the nineteenth century—was born. The practice of publicly disinfecting merchandise was also used for the first time, in the late fourteenth century in the Duchy of Milan.[12] Although ad hoc innovations, these techniques became permanent, even if they were less vigorously pursued during lulls between epidemics. For example, in 1400, Gian Galeazzo Visconti started a bureau of health officers, responsible for the preservation of health, which survived well beyond the end of his reign. Similarly (although not until 1486), Venice decided to replace temporary organizations, put in place at the time of successive dangers, with a permanent health commission.

The first two pillars of government sanitary regulation were inspection of travelers and quarantine of ships. The third was the lazaretto, an enclosed place for quarantining individuals, first instituted in Venice in 1423. One of the islands in the lagoon was designated for this purpose. This island was also the home of the Monastery of Holy Mary of Nazareth, which was dedicated to Saint Lazarus. In 1468, Saint Erasmus Island became a new quarantine center, run by the government bureaucracy's plague inspectors. The position of health commissioner was established in 1486. The responsibilities were vast, for this person was in charge of approving ships' health certifications, making decisions on quarantine, making sure goods were plague-free, administering the lazarettos, and ensuring the cleanliness of the city as well as its water cisterns, canals, food products, inns, and the homes of the poor. These pioneering measures did not develop in the rest of Europe until much later, in the seventeenth and eighteenth centuries. In Marseilles, officials assigned to health-related duties were mentioned in some documents in 1472–73, but permanent government health institutions were not established there until the sixteenth century. On the other hand, their leprosarium was converted to a plague hospital in 1476.

Faced with the threat of plague, it was governments, rather than the physicians at the famous Italian universities, that equipped themselves with the tools to fight epidemics. For several centuries, public health administration in the city-states of northern Italy was on the cutting edge. Moreover, their actions were not limited to the defense of their borders, as with such measures as inspections, quarantines, and lazarettos. Instead, their interests extended to urban sanitation and the cleanliness of foodstuffs. The authorities concerned themselves with promoting urban sanitation because of their belief in a theory we might call "aerism," which was a legacy of Arabic and Galenic medicine; according to this theory, bad air entered the body not only through the respiratory tract but also through dilated pores in the skin.

In the kingdom of France, slow to develop its health infrastructure compared to Italy, John II issued an edict, stemming from the plague of 1348, that applied throughout his kingdom. It required the cleaning of streets when there was a risk of an outbreak. In Reims two centuries later, a government order was promulgated to check the plague of 1544. It had five provisions. One concerned cleanliness of the streets, requiring that the gutters be washed out and cleared of their garbage. Another concerned the slaughter of animals, while a third allowed for the monitoring of beggars and required them either to work or to leave the city. A fourth required the participation of health care providers, barber-surgeons, nuns, and confessors to help deal with the epidemic. A final provision regulated the treatment of plague sufferers and their belongings. These patients had two choices: they could either be shut up in their houses for six weeks, or they could join a group of fellow-sufferers in isolation outside the city walls. The disinfection of houses, furniture, and clothing was required (although the process was limited to a good airing). The king imposed this rule through his representative. Two local worthies were to receive a list of plague victims every three days, drawn up using information from every parish priest. The penalties for breaking these rules were not specified.[13]

A different approach contributed to the spectacular successes of two major efforts to control the plague early in the modern era. First, in 1665–68, Colbert imposed restrictive measures involving the inspection of traffic entering Paris, a step that prevented plague casualties in that city from reaching the level seen in London, despite troop movements at the French border and in Flanders (due to the War of Devolution). Second, in 1720, the last occurrence of plague in Western Europe, which came in through Marseilles, was basically confined to Provence thanks to military encirclement of the region; the outbreak stopped early in 1722.

Cleanliness of the streets, the inspection of merchandise, and surveillance of foodstuffs had been realities in Italy from the fifteenth century onward. Public health policy, although certainly limited in scope, predated the hygienic reforms of the monarchies in the modern era, particularly in Italy.

Bad Air—or Planetary Misalignment?

Imported and spread by the Arabic scientists of Sicily and Spain, Greek medical knowledge was seen as new information. In the fourteenth century, Salerno, the foremost center for the study of Hippocratic medicine in the Middle Ages, was supplanted by Bologna. The faculty of medicine there had a curriculum based upon the work of Aristotle and Averroes and then the critical examination of the texts of Avicenna. The experience of the physician became part of something that could be taught as a doctrine, while being subject to minute scholarly examination, which was refined and improved (or perhaps worsened) by the critical analysis of the Arabic-Galenical texts that became part of the tradition. Medicine thereafter was both a science and an art, composed of both theory and practice. "Renaissance anatomy," based on the practice of dissection performed by a surgeon under the direction of a physician, arose at this time. However, neither the blossoming Italian universities nor their physicians, with their unbridled social prestige, provided the impetus for the establishment of the new public health institutions. No doubt the enormous accumulation of theory at that time was not conducive to practical advice. Still, the aerist doctrine undoubtedly should get credit for focusing the attention of officials on the disastrously unhygienic conditions that typified the cities of the period.

Another view was held by the professors at the Sorbonne. When asked by King Philippe VI, these learned physicians attributed the plague, even before seeing any cases, to a bad alignment of the planets. The Frenchman Guy de Chauliac, physician to Pope Clement VI in Avignon, adhered to the overall aerist doctrine. When the gravediggers of Avignon died, he had them replaced by mountain dwellers from plague-free regions nearby, figuring that fresh air was an effective protection against the disease. At the same time, however, he gave the pontiff good advice. Processions were suspended, the sick were moved to wooden cabins away from the city, and autopsies were conducted to determine causes of death. There was a striking gulf between the writings of learned professors of medicine, the actions of practicing physicians, and the beliefs of the population at large—after all, the people had always observed the contagious nature of numerous epidemics.

Flagellants and Pogroms

Witnessing the disappearance of half the population over the course of a few days is incredibly traumatic. Even observers accustomed to usually elevated mortality rates immediately take note: "Almost no one survived to the fourth day, and there was nothing to be done for it; no doctor or medicine was of any use. Whether it was a truly new illness or simply one that doctors had never studied before, it seemed that there was no cure for it."[14] Fear and anguish were followed by

flight, aggression, and mysticism. Flight was the most common recommendation of physicians, in accordance with Hippocratic and
Galenic advice on epidemics: "Flee immediately, to a great distance,
for a long time." But this suggestion was more easily followed by rich
people, who owned country places and had enough money to stay in
the country for an extended period. The poor, often lacking work because of the epidemic and driven from the villages they wanted to
pass through, were frequently forced to stay in the cities. Sometimes
people sought protection by trying to flee to the house of a physician
or a charlatan or a crank; others sought holy miracle workers. Finally, some could not stand the ongoing anguish generated by the
plague and could not endure the death of so many of their loved
ones; they killed themselves.

Some people reacted to the plague by joining the ranks of peni-
tents called *flagellants* (who had become more numerous in Italy
since the mid-thirteenth century), first in Sicily at the start of the
epidemic and then in Venice from August 1348 onward. Flagellants
appeared next in Vienna, Hungary, Poland, the German states, Flanders, and then finally in France. The chronicler Albert of Strasbourg
reported that, in 1349,

> they came from Swabia to Speyer in mid-June, led by a chief and two su-
> periors. When they had crossed the Rhine, at one in the afternoon, a
> crowd of people rushed up to them. The flagellants formed a big circle in
> front of the town monastery, and some of them got undressed, wearing
> only a covering running from the small of the back to the heels. Thus at-
> tired, they walked in single file around the circle, with their arms ex-
> tended in the form of a cross. Each one then lay face down on the ground
> while the others stood over them, with one leg on each side of the person
> on the ground. Each person on the ground received a little blow from a
> whip; they all got up and then proceeded to beat themselves with whips
> on which there were knots and sharp pieces of iron, while they sang
> psalms . . . There were about a hundred men from Speyer and about a
> thousand from Strasbourg who entered into this brotherhood and swore
> obedience to the superiors for the allotted time. To join, they had to have
> four coins per day in spending money, so as not to be reduced to begging.
> They also had to declare that they had confessed, that they were contrite,

that they had pardoned all their enemies, and that they had the consent of their respective wives. Finally, they left Strasbourg in two groups, departing in opposite directions. One went down the Rhine, and the other went back up. An enormous crowd then ran along with them.[15]

The movement reached its peak at Christmastime in 1349.

The religious and civil authorities condemned the flagellant movement and attempted to break up the groups of flagellants, which by then formed a genuine international mystical sect whose followers committed themselves to joining the pilgrimage for thirty-three and a half days. They believed that this penance regained for them the purity of their baptism, and they reasoned that, as a result, they would be immune to illness and the plague. This outburst of collective hysteria was revived from time to time, as in Provence in 1398 and in Paris in 1583. At the end of the sixteenth century, the flagellant movement was prohibited, and this form of public ritual expression disappeared.

Collective aggression was usually directed toward marginalized populations that were accused of spreading the disease: the poor, lepers, and the Jews. Boccaccio was not the first to have noticed that the disease first arose in poor neighborhoods and that it was particularly deadly there. Faced with this "biological" peril, cities sought to protect themselves by removing the poor. At Bourg-en-Bresse in 1472, not only beggars but all poor people were kicked out. Identical measures were taken in Troyes in 1517, in Amiens in 1545, and in Agen in 1605. "Poor-hunters" were employed at the gates of very many cities; it was their job to see that beggars did not get in.

In Switzerland and Germany, massacres of Jews preceded or accompanied the arrival of the flagellants, who sometimes also brought the plague. Violence against the Jews was heightened by the epidemic but had occurred before it. In Aquitaine, early in 1320, rumors had already accused Jews and lepers of poisoning the fountains and wells to spread diseases. At Chinon, lepers were massacred. In June 1321 several were executed after having "confessed" to poisonings (for example, in Tours and Perigueux). When the plague spread rapidly in the spring of 1348, animosity was directed against the same groups. In April, forty Jews in Toulon were massacred and their homes looted, while at Narbonne it was the poor who were accused of bringing poisons and who were condemned to death. In contrast, lepers seem to have been forgotten, and, very rapidly, mob violence came to be focused exclusively on the Jews. This anti-Semitism may be explained by the way of life forced on the Jews, who were required to live in communities concentrated in ghettos and who were not allowed to own property. Their small money-lending businesses were resented by a multitude of short-term debtors. The authorities tried to protect them—Pope Clement even issued a bull on their behalf—but, overwhelmed by numbers, the authorities often gave up under the pressure of the mob. Half the Jews of Strasbourg were burned to death, and the rest were forbidden to be in the city at night. They were also killed off in numerous towns in the German states, such as Basel, Fribourg, Ulm, Baden, Constance, Dresden, and many others.

The Danse Macabre and the Apocalypse

Omnipresent sudden death provoked diametrically opposed reactions. On the one hand, there was a sort of passion for life among the peasantry, seen in the growth of festivities and entertainment in the cities, and in carefree, sardonic attitudes, not to mention the growing splendor of life at the royal court. On the other hand, a sort of macabre exhibitionism arose in the fifteenth century. Representations of decomposing bodies were sculpted on tombs, for example, that of the famous Cardinal Jean de Lagrange at the Church of Saint-Martial of Avignon. Many decorative objects bore numerous grimacing skeletons, including frescoes portraying macabre scenes of dead, decaying, or skeletal people dancing wildly. Moreover, after irreducibly minimal funeral ceremonies, the dead were buried in collective tombs. Death, stripped of its rituals, was no longer humanized by rites, tears, and prayer. According to a chronicler from Sienna, Agnolo di Tura, "No church bells rang, no one cried . . . And people said and believed: 'It's the end of the world.'" G. Cosmacini emphasized that, at that time, "the grave became a depersonalized mass grave from which a crowd of anonymous corpses would emerge on Judgment Day." The face of death was different; it had become a sort of pagan goddess whirling in the winds, a profane being, a living aspect of the dead body of the plague victim. Death was seen no longer in the context of Christian providence but, instead, in the context of a naturalistic view; death was lacking in purpose and was simply the incarnation of the illness that preceded it.

Artists tried to deal with fear by using a form of macabre exhibitionism. In frescoes and miniatures, the motif of death leading a dance arose in the fifteenth century, as in *Les Heures de Paris*. The paintings in the galleries of the Cemetery of the Innocents, in Paris, were the first to feature this theme. They date to 1424; many more were to follow.

The depiction of the dance of death often took the form of a medieval dance called the fandarole, and in the center of the group of dancers one might find cardinals, kings, and emperors just as easily as one might find ordinary priests, physicians, and simple peasants. Death made no distinction among wealth and social levels, any more than it distinguished among age groups. These depictions of cheerless dances have often been interpreted as a reminder to all, whether powerful or lowly, of the fragility of life. Many cloisters and churches were decorated with painted pictures of a danse macabre. Examples include la Sainte-Chapelle de Dijon, 1436; La Chaise-Dieu, prior to 1460; Basel's Dominican convent, around 1440; the Marienkirche in Lübeck, 1463; the Marienkirche in Berlin; and churches in Estonia and Finland. Danse macabre decorations were very rarely found in England (although the earlier Saint Paul's Cathedral had one example), and they were found in Italy slightly later (at Clusone in 1485, Como in 1490–1500, and Pisogne in 1490). No doubt because of their popularity from the late fifteen century onward, this type of image was carried over into the first woodblock prints and inspired more frescoes in the sixteenth century. There were wooden sculptures of the danse macabre, too, such as those at the Saint-Maclou ossuary in Rouen and the Saint Saturnin cemetery in Blois. Artists abandoned this theme during the Renaissance, although it resurfaced occasionally in literature and music.

At the time of the plague, the theme of the Apocalypse also became a source of fascination and inspiration. It was evoked with new vigor in the Angers tapestry, created around 1380. Another fine example is the painting by Fra Angelico and Benozzo Gozzoli, in the San

Danse macabre: After the Great Plague, death is represented as the great social leveler. Vagrants and poor people are in the same dance as bishops and wealthy people, men as women.

From *Bernt Notke und sein Kreis,* by Walter Paatz (Berlin: Deutscher Verein für Kunstwissenschaft, 1939).

Brizio chapel in the Cathedral of Orvieto. It shows the preaching of the Antichrist, the end of the world, the resurrection of all humanity, and the Last Judgment.

The church stopped tolerating "white magic," like belief in enchanted trees or springs with curative powers for humans as well as cattle. Traditional, harmless practices involving these sources were now perceived as dangerous by authorities, who saw evil everywhere, and practitioners were persecuted. Starting in the early fifteenth century, there was a huge increase in the number of people of both sexes who were tried for witchcraft. Men and women were certainly expected to communicate with God and the saints, but only in approved ways—through prayer, processions, veneration of relics, and pilgrimages. At that time, a great many religious practices aimed at protecting followers from various diseases arose or expanded. Saint Anthony, Saint Roch, and Saint Sebastian were invoked against epidemics, Saint Christopher against sudden death. Against misfortune, one could invoke the entire heavenly host, especially the sorrowful Virgin of the Pietà, not to mention Saint Catherine or Saint Margaret.

With the large-scale dislocation and disruption of families, individualism increased. This led to a marked expansion of private prayers and the celebration of thousands of masses for the dead. The fourteenth and fifteenth centuries were the golden age of brotherhoods, which were a kind of welcoming network and mutual aid society. The epidemics of the time contributed to the awareness of the fragility of life and to the desire to prepare for one's death and one's salvation. This awareness is evident in the meticulous arrangements specified in wills. The mendicant orders, the Franciscans and Dominicans, were very successful at the time, having become indispensable in the consolation and care of the dying.

Plague outbreaks punctuated life in the French kingdom every eleven or twelve years from 1347 to 1536 and every fifteen years from 1536 to 1670. Sometimes the epidemics were regional, but often they affected the entire kingdom. Over three centuries, twenty-six major surges of plague mortality have been identified, to which must be added eleven secondary increases. The epidemics greatly destabilized economic and social life, yet they also provided opportunities to establish the first local policies aimed at protecting the public's health, especially in the large Italian cities, which established new institutions and practices that would have a long history. The medical establishment, demoralized by the first outbreaks, later tried to change practices in order to forestall new epidemics. Finally, the church took advantage of the disarray in society to impose its rituals, strengthen its religious centers, and compel behavior in accordance with its views, using the threat of death by being burned at the stake.

New Concepts of the State and the Body

Plague receded beginning in the middle of the seventeenth century, and it disappeared from Western Europe after the last outbreak in 1720. This last epidemic was violent, seemingly an echo from a vanished past; however, it was limited to Marseilles and Provence. Possibly the replacement of one variety of rats by another played a role as plague receded. An undeniably new factor was the effectiveness of the steps taken by the royal governments and of the coercive measures that managed to block the geographic expansion of the disease. In addition, new attitudes and practices related to the body resulted in an explosion of attention to the care of the body, at least among the elites and provincial notables. In the field of medicine, the outlook also changed, resulting in the hygienic movement of the nineteenth century. The interaction of all these factors was the necessary precursor for the first step toward the control of epidemics.

Economies of Scale

The story of plague hospitals, quarantines, and restricted travel is the story of local authorities' struggle against the invasions of plague. It is, however, also a story of bungling. Despite their long-recognized legitimacy, the public health institutions of the Italian cities encountered much opposition until the middle of the seventeenth century. A telling example is provided by the well-documented implementation of sanitary regulations in Prato, a Tuscan town, during the plague threat of 1629.[1] Despite strict prohibitions on travel, surveillance at the city gates, and the implementation of health passports, plague entered the town in August 1630. Local health authorities requested the permission of their superiors at the Florence Board of Health to build a lazaretto. To avoid panicking the people, however, and out of concern that other governments would suspend contact with all of Tuscany, their superiors did not grant permission. Eventually, when it became undeniable that there really was plague in the town, a site was sought for the lazaretto. It was still necessary to reach a settlement concerning the villa in question and to overcome the resistance of the owners, who had won judgments in their favor twice in Florence. The commissioner of health did not have the ultimate power to implement necessary measures and turn them into practical reality. When confronted with well-off families, he had to compromise. Moreover, because of a lack of manpower, he could not stop the townspeople from going to a neighboring village for the grape harvest (even though he had forbidden it due to the plague), any more than he could prevent the sick people shut away in the lazaretto from escaping! The powers of the health commissioner, as extensive as they seemed, were limited in practice by the interplay of social forces, at least at the start of the seventeenth century.

In France, too (later than in Italy), towns organized checkpoints and systems to isolate themselves in case of danger. In Reims, for example, which suffered frequent waves of plague starting in the fourteenth century, inspectors were put in place at the town's five gates as soon as attacks occurred in Paris or the surrounding areas. In fact, in June 1619 the town council kept the inspectors in place despite the king's order to remove them and also made an effort to eject all beggars and vagrants to outside the town walls. A "bureau of health" was established at the town hall to provide overall coordination of sanitary activities. The focus of these activities was control of the gates of the city by the local militia, in order to deny entry to people or goods coming from the capital or other infected towns.[2] Furthermore, the inhabitants of Reims were forbidden to have as a visitor or an overnight guest anyone lacking a certificate of good health issued by officials in the visitor's hometown. Every evening, owners of hotels and taverns were required to give the bureau of health a list of their customers, showing where they had come from. Although lapses were punishable by fines, infractions were numerous. Churchmen returning from Paris made people allow them back in, children of distinguished families were allowed to return home, and shop-keepers bought bundles of wool from infected regions in defiance of the ban. Every regulation was circumvented as a result of various social and business interests. The fines imposed on the entire Reims citizenry did not exceed a total of eight livres, which was scarcely the equivalent of two weeks' salary for a stonemason.

Only the intervention of the French royal state could make effective implementation a reality. Compared to merely local authorities, the royal state controlled territory and resources on a much larger scale. Two famous examples provide perfect illustrations of the improved implementation of public health measures made possible by an increased scale of operations. Plague returned to Amsterdam late in 1663 and then spread in the Netherlands and in the British Isles. The parlements of Rouen and Paris put controls in place immediately, but Lille was hit in November 1667; during the following spring and summer, Amiens, Laon, Beauvais, Reims, Rouen, Le Havre, Dieppe, and many neighboring areas were also affected. The epidemic receded during the winter of 1668 and disappeared totally by the beginning of 1670. What happened? Thanks to determined measures by the central government, Paris was able to escape the plague, in contrast to London, which was devastated. Jean-Baptiste Colbert put back in force the preventive measures that were standard for such situations, of course, but in an important change from the past, he made sure that they were effectively applied despite numerous obstacles. Turenne was leading the War of Devolution in the Netherlands, and he located his quarters there in the winter of 1667, which would require frequent troop movements in the area. Also, entire regions—the Île-de-France, Picardy, Normandy, Champagne—were dependent on manufacturing and trade. Despite the many pressures from representatives of special interests, Colbert kept in force the rules that forbade traffic there. Quarantines on goods and people crippled economic activity and consumed resources. Colbert had to choose between economic activity (a great concern of his) and the protection of the health of the capital and the surrounding region. He chose the latter and succeeded in confining the plague to the northernmost regions of France.

The second example of the newly effective coercive measures is during the last plague epidemic in France, which occurred in Marseilles in 1720–21 and which is extremely well documented in both primary sources and secondary works.[3] Plague first appeared within the city walls on July 22. Communities such as the Arsenal, the Abbey of Saint Victor, Fort Saint-Jean, and the walled Citadel of Saint Nicholas cut themselves off from the city. On July 31, three thousand beggars from other towns were expelled, while local beggars were locked up, and the same day the parlement at Aix required that Marseilles be sealed off. But it was too late: more than ten thousand people had already left and spread the illness to the surrounding area. Ports and bridges on the nearby Durance River were then closed.

The boundaries of the isolation zone, promulgated on August 4, weren't really secure until new troops arrived on August 20 and completely blocked off the city and other affected areas. Business was conducted within shouting range of four "barrier markets," where goods brought from the unaffected zone were left on the road, in a sort of buffer zone; the people stuck in the blockaded city could then retrieve them. The towns of Provence shut themselves off from the outside world and forbade entry to any traveler without a health certificate. They even organized blockades, such as one running the length of the canal at Craponne and another on the river Siagne. For its part, the provincial government set up military cordons they called "lines" on the Verdon, the Durance, and the Jabron as far as Buis-les-Baronnies. The Durance was guarded by the government of France and by the region of Comtat, which was the possession of the Holy See. A new line was established between the Comtat and Haute-Provence on October 2; then it was decided to make it insurmountable, a goal achieved by a deep, wide trench topped by a wall of sharp stones. This "plague wall," sixty miles long, is still visible today; at the time, it was built very quickly and involved a significant financial burden. But it did not manage to protect Avignon, so it was necessary to construct another line farther out from the wall in February 1721. This wall was abandoned on September 22 because it had become useless and financially ruinous. The following year, three battalions of French soldiers were sent to reoccupy the line to protect Provence, then plague-free, from any possible reinfection. They did not with-draw until February 1723, three months after the end of the epidemic at Avignon.

In Languedoc, too, starting in August 1720, the big towns estab-lished health departments and attempted to control the flow of peo-ple. A military blockade was established, extending the length of the Rhône from Viviers to the delta and along the stretch from Saintes-Maries-de-la-Mer to Leucate. Gévaudan was struck by plague start-ing in November or December 1720, so the following spring it was decided to establish a new line at the boundary between Guyenne and the Auvergne and extending the length of the river Orb. On two occasions, these lines had to be moved westward and northward in response to the advancing epidemic. Not counting municipal inter-ventions, the region of Languedoc was maintaining two thousand guard posts until December 1722.

The measures taken were as dramatic as they were costly. Despite the stiff penalties for anyone crossing a line or disobeying plague-re-lated rules—and there are examples of people being shot—the epi-demic did break through the line. This was sometimes the result of carelessness, as with the guards who played cards with the suppos-edly isolated people of La Canourgue. It was sometimes the result of crime (shady characters who were fleeing or smugglers, for exam-ple). Sometimes it was the result of the migrations of seasonal travel-ers who helped gather the crops from farms or vineyards and man-aged to reach the locations where traditional seasonal labor was needed.

This final plague, ultimately restricted to a large region in Pro-vence and a part of Languedoc, evoked the biggest deployment of sanitary blockades against the plague ever seen in France. The block-ades extended over very large distances, were reset depending upon the movements of the epidemic, and were supplemented by addi-tional lines as needed. The objective was, naturally, to keep the epi-demic isolated in the countryside and prevent it from spreading to the towns and their surrounding areas. Such practices originated in Spain in the middle of the seventeenth century, when the goal was similarly to isolate entire infected regions using an uninterrupted blockade. Only large kingdoms in the modern era were capable of imposing and financing such measures. In strengthening them, they succeeded in implementing effective control measures against plague.

The principal factor in the successes was certainly the change of scale involved: the goal was to seal off the entire countryside, not a town, even if that meant increasing the number of cordons, using draconian means to enforce them, and maintaining them for several years if the danger remained. This kind of concerted protective action against the plague involved high costs that only the state could afford.

This forceful policy can be seen as a consequence of mercantilism and, if one follows Michel Foucault, as a manifestation of a new policy: control of people's lives by the state. Starting in the seventeenth century, this control over the life of the individual developed on two foundations. The first involved the concept of the body as a machine. Training and adherence to specific instructions would promote the care and training of the body, increase the body's abilities, and optimize the application of its strength. The second means of social control involved thinking about society at large as if it were a single organism possessing biological characteristics, such as "rate of proliferation, births and deaths, health status, and longevity, subject to the circumstances that can cause these measures to vary. A hold on power is made possible by a whole series of measures and regulatory controls, which can be termed a 'bio-political policy' of the population," as Michel Foucault writes. For the first time in history, biology was reflected in policy; living itself became part of the domain to be controlled by knowledge and power. "We use the term 'bio-policy' to designate anything that subjects life and its mechanisms to calculation, anything that makes 'know-how' a means of transforming human life." We might speak of a society "crossing the threshold of biological modernity" at the point where the population is betting the species on the success of its own strategies. "Modern man became a political animal from the moment when his existence as a living creature was placed in question," Foucault says.[4] At that point, too, there arises in consequence a new way of conceptualizing the body.

The Care of the Body

Starting in the seventeenth century, the large European monarchies considered their populations to be the source of their power and wealth. Georges Vigarello clearly shows that this attitude went hand in hand with the new views of the body.[5] After the sixteenth century, the body seemed less dependent on invisible forces. The Protestant Reformation criticized the veneration of relics, and it now seemed possible to evade the cosmic forces of destiny by, for example, following a good diet. Much more attention was paid to rational principles and to the connection between cause and effect. The body was now compared to mechanical, manufactured objects. For example, the body is like a still that purifies and distills the humors within it; drinking "firewater" extends the analogy. At other times, the body was seen as analogous to a republic, or a town, or even the army, with each organ contributing to the functioning and unity of the whole. All of these representations are based upon a view of the body as a physical unity that is orderly and hierarchical, with the periphery subordinate to the commands of the center. Accordingly, it became common to associate "maintenance of health" with "government" and to connect "behaviors that protect health" with "orders." The seventeenth century saw the development of concepts of the body as a system of piping, flows, and pumps; a little later the body was like a machine with pulleys and levers. For a long time these concepts were restricted to small elites: those in medicine, the aristocracy, and the bourgeoisie. The older concepts of the body were deeply rooted in the general population, which was shown dramatically by Enlightenment physicians' ceaseless attacks on popular beliefs, prejudices, and practices, which they pursued in part to improve standards of behavior. For example, Louis Lépecq de La Cloture, a physician responsible for epidemic control in Normandy and thus a member of the elite, deplored the use of "spiritous liquors" by the peasantry as a treatment for intestinal ailments.[6]

Many questions remained beyond analysis and understanding. No doubt the Renaissance rediscovery of antiquity's aesthetic view of the body contributed to the cult of youthful beauty, an emphasis on

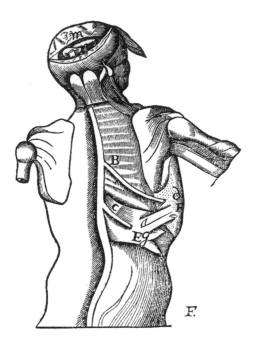

Descartes's knowledge about how the human body works is evident in this illustration of muscles, bones and joints, and the nerves as a complete system.

From *Treatise of Man,* by René Descartes.

the pleasures of the here and now, and a renewed interest in the human form. This development took place simultaneously with an aloofness from religion, which was marked at the time by wars between Catholics and Protestants and by worsening theological quarrels. A system of thought not based on strictly religious grounds thus developed. One can readily understand the reciprocal influences between the urban upper crust and learned physicians, who associated with each other. It is more difficult to understand the spread of new concepts among rural doctors and the provincial higher classes, unless one takes into account the means of spreading these new ideas, namely, the social circulation of city dwellers and the growth of academies, associations, and philosophical and scientific societies, which began in earnest at the end of the seventeenth century.[7] The

Royal Society of Medicine certainly did not have any difficulty con-
tacting its members and correspondents.

On the other hand, the diffusion of the new ideas among the common people was more complex, slower, and also less dramatic.
François Lebrun points out that, until the middle of the eighteenth
century, one could not really separate scientific or natural medicine,
on the one hand, and the parallel system involving magical or super-
natural power, on the other. Doctors in the medical elites, like the
empirics, used "natural" means to restore health, while those who
turned to exorcists and healing saints called upon only magical and
religious forces.[8] But knowledge circulated from one place to the
other, and the same patient used techniques that we today would
consider to be in drastically different, or even opposing, categories;
even among the elites, some employed both empirics and magical
cures.

Among the common people, visits to a doctor were necessarily
limited by economic means; their views of the body, of illness, and of
physicians have left but a few traces. According to Françoise Loux
and Philippe Richard, one can try to find clues in proverbs, which
were abundant in the health field.[9] Proverbs also express the terror
provoked by illness, the ultimate scourge, and the solace of religion
("It's faith that cures") and of the saints ("Saint Blaise can assuage
any illness"). We are also told of the power of natural remedies
("Mint washes away all illness") and encouraged to doubt doctors
("When it comes to fever and gout, doctors know nothing"; "Doctors
and cavalrymen kill people and horses").

A genre of literature for the masses, sold by peddlers, provides
another insight into popular attitudes toward illness. *The Friendly
Apothecary*, for example, details medicines that can be produced
quickly and cheaply at home. These works occupy a middle ground
between the learned literature (they did borrow from the medical
knowledge of their day) and common practices. Dom Nicholas Alexandre, author of the 1714 book *Medicine and Surgery among the Poor*,
states at the outset that the effectiveness of remedies is not proportional to their cost:

It is to be expected that drugs made from ordinary and even disgusting ingredients would be distrusted and rejected by the rich, who value only those with rare, costly ingredients brought from the Indies. However, very often the effect of these latter remedies is to empty the purse without restoring health, while the common folk cure themselves of the same illnesses promptly and perfectly well using simple, familiar remedies . . . These get termed "old wives remedies" by those who want to make people distrust them.

The specific powers attributed to particular saints with curative abilities were still very important and often pertained to some tragic event in the saints' lives. Saint Lawrence, for example, who was burned to death, cured burns; Saint Apollonia, whose executioner removed her teeth, cured toothache; Saint Vincent was disemboweled without showing any sign of pain, and he became the patron saint of those with intestinal ailments. Saint Agatha's breasts were cut off, and she cured problems with inadequacies of mothers' milk. Saint Sebastian and Saint Roch were especially invoked in times of plague, the former because of the similarity of the arrow wounds of his martyrdom to the buboes of the plague patient, the latter because he was miraculously cured of a terrible illness. Sometimes a play on the name was enough: for example, Saint Méen was said to be able to cure problems with the skin on the hands [in French, the word for "hand," *main,* is pronounced much like this saint's name]. Saint Claire was invoked for eye ailments interfering with clear vision, and Saint Aurelian for hearing (aural) problems.

The situation, however, was not as rigid as the above discussion might lead you to believe. And for a long time, traditional attitudes coexisted with an openness toward novelty.

A Cure at Any Cost

The concern with protecting the body and the use of new products to achieve this goal date from the early eighteenth century. The demand for care and medications probably predates the supply. At this time, publications on medicine aimed at the general population went through numerous editions. *An Anthology of Easy Home Remedies* by Madame Fouquet had sixteen editions between 1675 and 1740; *Medi-*

cine and Surgery among the Poor by Dom Alexander appeared in four editions between 1714 and 1758. New products, notably cinchona bark and ipecac, introduced to Europe in the seventeenth century, enjoyed a popularity that quickly spread beyond the aristocracy. The market proved to be so lucrative that it provoked speculation on the part of investors. Helvétius built a part of his fortune on the commercial exploitation of ipecac. The consumption of mineral waters, such as those from Vichy, Plombières, Forges, Barèges, and Bourbon-l'Archambault, also increased greatly in the eighteenth century, as did the sales of patent medicines. Inventors sought official recognition for their patent medicines, and they were sold by peddlers in the most obscure corners of the countryside. The rise of patent medicines was not the consequence of a wave of ignorance but rather an indication of a "new quest for a cure at any price." For this reason, the era has been referred to as "the golden age of traveling salesmen."[10]

The growth in the numbers of physicians and surgeons is also evidence of increasing demand. In Caen between 1750 and 1792, medical practitioners increased in number by 39 percent, while the town's population grew only 9 percent. In Paris, the number of surgeons nearly doubled, from 235 in 1715 to 466 in 1789; this increase, too, was well in excess of population growth. Alain Corbin has provided scholarly evidence of a revolution for body cleanliness and against body odor, which began in the second half of the eighteenth century; he also documented how slowly this revolution spread throughout the general population.[11] Attitudes toward hygiene and the practice of hygiene were transformed, in both public and private spheres, and these new attitudes created circumstances conducive to additional improvements.

The Decline of Mortality

Several studies seem to show an initial decrease in mortality at the end of the seventeenth century and the start of the eighteenth. At Tourouvre-au-Perche, for example, comparing the periods 1670–1719 and 1720–69, the mortality rate among children under age five decreased by one-fourth. Around 1690 in Geneva, there was a decline in the average mortality for children under ten, one of a series of declines resulting in ever-lower plateaus of mortality. Life expectancy at birth for those born between 1659 and 1699 was two and one-half years greater than it had been for those born from 1600 to 1649, and this result was primarily due to declining mortality in the first few years of life. A study of the Benedictines of Saint-Maure provided evidence of an improvement of life expectancy at twenty-five years of age between the period 1624–74 and the period 1675–1725.[12]

This initial diminution of mortality—a transition occurring over the course of two centuries—was followed by another decline, a century later, at the time of the French Revolution and the Empire. Male infant mortality was 291 per thousand in 1770–89 but was only 212 per thousand in 1810–19, an improvement of about 33 percent. Thereafter, progress was not limited to children: the chance of reaching age sixty went up by more than 40 percent between 1780–89 and 1800–1819.[13] Yet improvement was most pronounced among the babies of rich women. Considering parents who married in Geneva between 1625 and 1684, the death rate among the children of the bourgeoisie was half the rate among children of laborers and unskilled workers. Marked variation between social strata was found in Rouen, too, and was also certainly present in the countryside, where the contrast was substantial between the wealthiest people of the villages, who were not at risk of malnutrition, and the far larger group of peasants, including those laborers who were at the highest risk.[14]

The disappearance of the great epidemics of plague after 1670 (except the episode in Marseilles from 1720 to 1722) was the decisive factor in these two successive mortality declines. This is true even though other disease scourges became more important. It has been estimated that the plague killed 2.4 to 3 million people in France during the seventeenth century. According to Alfred Perrenoud, smallpox took 5 to 6 million lives, and typhus, dysentery, and various fevers were omnipresent.[15] Climate fluctuation does not explain the improvements, for when such fluctuations reversed course, the gains that had been achieved were not lost. All the evidence suggests that several factors were responsible for lessened mortality. A decline in disease outbreaks was helpful, to be sure, but so were improved living standards, new attitudes toward the body, and new collective efforts aimed at protecting health.

From Helvétius to Vicq d'Azyr

Changed perceptions of the body and of the causes of disease went hand in hand with mercantilism and the effects of Jean Bodin's famous statement, "All wealth and force derive from the people." New perceptions contributed to the establishment of an unprecedented public health network in France, with responsibility residing in provincial administrators. Starting in 1710, these administrators received free boxes of medications annually, prepared according to the recommendations of Helvétius; the shipments were to contain the remedies that would be most useful for curing the currently most prevalent diseases. In 1750, a physician concerned with epidemics was appointed to each province, with responsibility for handling any outbreaks of disease anywhere in the province. Clearly, there was a hierarchical, pyramidal organization to this governmental effort.

The creation of the French Royal Society of Medicine in 1776–78 marked the culmination of an effort begun in the 1730s and aimed at establishing a network of twenty-four physicians in the provinces; this network would facilitate the establishment of a true sanitary inspection system as well as more coordinated—hence, more effective —sanitary measures. By 1732, this initiative had disappeared, along with its originator, Pierre Chirac, the first physician of Louis XV. When it was taken up again thirty years later, the purposes were to improve both the dissemination of information about medical problems and control over the spread of disease. The Royal Society of Medicine, which enjoyed the support of the king and the participation of many young provincial physicians, made a special study of questions relating to health and public hygiene. From the moment it was founded under the impetus of Vicq d'Azyr, it organized the collection of a wide assortment of observations from meteorological to medical data. Meteorological and climatic data were considered important because of contemporary concepts about the relationship of the body to the environment. These concepts demonstrate a strong neo-Hippocratic influence, which held that illnesses arise in specific environmental conditions; thus, illness could be prevented by altering a single environmental element. This principle was not limited to the physical environment but extended to the effects that might be obtained by changing the habits of the population. For the first time in France, an organized and meticulously hierarchical system was put in place in the capital.

Around the same time, projects to develop public health legislation arose. One of the best known is the set of initiatives put forth by Johann Peter Frank. He was inspired by the medical police measures taken in France in the early 1700s and compiled by Nicolas de la Mare or Fréminville. The new rules regulated the purity of water, the quality of food, drinks, and medicines, and the health and safety of children under the care of wet nurses. They even regulated the maintenance of latrines and the practice of medicine. Frank was an adherent of the theories of Rousseau, according to whom social organization and inequality were the main sources of illness. Humans, naturally healthy, disturb natural balances by their behaviors, but nature fights back with disease. The main job of medicine is to reestablish conformity with the laws of nature when they have been violated by "excessive passions" and by the luxuries with which people have learned to surround themselves. Inequality breeds poverty, and poverty breeds diseases. Medicine's potential is limited by these factors. Thus, physicians have the responsibility to convince princes— the powers that be—of these principles. The goal is to establish medical and public health laws that would prevent most, or at least the most serious, violations of nature,[16] controlling violence, protecting the public against contagious diseases, and regulating marriages to avoid increasing the number of cases of hereditary illnesses.

Fresh Air and Clean Water

The decline of mortality in the eighteenth century was in part the result of changes in agriculture, such as more efficient and less wasteful shipment of wheat and the substitution of potatoes for poorly preserved barley. But the mortality reduction was also among the first fruits of the hygienic reform movement.[17] In the late seventeenth and early eighteenth centuries, French physicians and eminent colleagues from neighboring countries (such as Thomas Sydenham in England, Bernardino Ramazzini and Giovanni Lancisi in Italy, and Friedrich Hoffmann in Germany) proposed specific interventions based on the Hippocratic tradition. According to this tradition, usual (endemic) illnesses, as well as less common (epidemic) ones, arose due to environmental circumstances. The hygienists prescribed the draining of marshes, bogs, drainage ditches, moats, and all bodies of stagnant water. Water in canals had to be moving to be usable for cleaning houses, and fresh air had to circulate in homes and in all places where people gather. And it was even advisable to burn sulfur, among other measures, to control insects in homes, prisons, hospitals, and boats. A new obsession took hold, focused on getting rid of bad air produced both by decaying matter and by the exhalations of human beings. Numerous systems of ventilation, some of them quite sophisticated, were invented. Beginning in the 1740s, removing bad air was among the recommendations preferred by physicians and health authorities in Western Europe. According to James Riley, practices like using water to remove town refuse, organizing garbage collection, and controlling stagnant water must have had a considerable effect on the populations of insects, although they did contribute to pollution of the rivers. In France, the many drainage projects dating from the end of the seventeenth century reduced the toll from malaria. Similar projects were undertaken in England, in the German principalities, and in Italy. Insects play an essential role in the transmission of dysentery, typhoid, typhus, and malaria. Below a certain threshold of density, insects would be the weak link in the chain of transmission, which might then be interrupted. Changing attitudes toward waste, garbage, cleanliness, and stagnant air and water clearly were fundamental to improving health beginning in the eighteenth century.

Vaccination and the Elites

A decline in fatalism is evident from the increasing demand for physicians' services among the elites (who waited with dissatisfied impatience, starting in the seventeenth century, for progress that was

slow to come, as evidenced in Molière's plays). This decline in fatalism is also shown in the Western European importation of the practice of inoculation against smallpox.

In April 1721, Lady Mary Wortley Montagu, wife of the British
ambassador to Turkey, had her daughter inoculated in London in the
presence of several physicians of the court of King George I. Inoculation involved taking pus from a smallpox patient's pustule, which
contained the smallpox virus and, by scarification, introducing the
pus into the body of a child who had never experienced the illness.
Lady Montagu had already tested the benefits of this technique by
having her son undergo the procedure at the hands of a Greek female
inoculator in Constantinople. Traditionally, the remarkable demonstration of 1721 is credited with marking the start of inoculation in
Europe.[18] Quite apart from the imitation of enlightened aristocrats,
however, a sort of "folk inoculation" had long been practiced in villages and the countryside in Auvergne and Périgord, not to mention
Corsica, Wales, Scotland, Denmark, and various regions of Italy, the
Mideast, and Africa. The widespread adoption of the practice in England was due to the willingness of the Princess of Wales to have her
own children inoculated in April 1722. The greatest doctors and surgeons then inoculated more than two hundred people from all walks
of life.

Inoculation was sometimes fatal, of course. The risk of dying as a
result of the procedure was perhaps 1 in 53 or even as much as 1 in 50,
according to the results of a survey reported by the inoculation prac-
titioner James Jurin in *An Account of the Success of Inoculating the Smallpox in Great Britain*, first published in 1723 (with several new editions
by 1729 and a French translation in 1725). Still, the risk of the inoculation was much smaller than the risk of the disease. In France at the
time, to Voltaire's amazement, smallpox inoculation was still at the
stage of preliminary theoretical discussions, and he, too, saw the issue in terms of relative risks. "Don't the French like living? Don't
their women value their beauty? . . . Out of every one hundred people, at least sixty will get smallpox. Of these sixty, in a year when the
disease is mild, ten will die and ten will bear the marks of the disease
for the rest of their lives. Thus, it is a certainty that at least a fifth of

people get killed or disfigured by this disease. Of all the people inoculated in Turkey or in England, no one died, other than people who were weakened to begin with or who died of other causes."[19] Voltaire's last statement was borrowed directly from Jurin, who offered those explanations for the 9 deaths among 481 inoculated in Turkey and the 17 deaths among the 847 inoculated in England. In England, the practice enjoyed such an increase in the 1740s that several tens of thousands of people were inoculated. In 1748, the Geneva physician Théodore Tronchin, who wrote the articles on inoculation and smallpox in Diderot's *Grande Encyclopédie,* inoculated his eldest son as well as the children of the duc d'Orléans. Geneva and certain Italian regions adopted the new preventive measure. Despite the impetus provided by Turgot, the example set by some in high social strata, and a remarkable success in the region of Franche-Comté, the number vaccinated throughout all of France between 1760 and 1787 must have remained below sixty thousand people. The practice of inoculation was introduced much more readily in Denmark, Norway, and Sweden.

Nevertheless, the new movement among the elites is a strong indication of their desire to take action when faced with the danger of this epidemic disease. In some English shires and in certain areas in Sweden, the severity of smallpox epidemics in the second half of the eighteenth century was certainly reduced. There is a threshold effect, and we do not know what proportion of children under five years old must be inoculated for herd immunity to prevent the disease from becoming endemic and to curtail epidemic outbreaks.

Vaccination's Astonishing Success

The famous story of vaccination against smallpox began with "variolation"—the transfer of infectious smallpox material from the sick to the healthy for preventive purposes—which had been in use for several centuries on several continents. The variolation process had been introduced to Western Europe by Lady Montagu with much fanfare, but it had limited success. The opposite was true of the procedure subsequently discovered by Edward Jenner, which spread rapidly.

Here's the story: Jenner's essay, *An Inquiry into the Causes and Effects of Variolae Vaccinae*, was published in London in October 1798. Odier wrote a review of it in Geneva, in the *Bibliothèque Britannique* (1799). Jenner had noticed that milkmaids in contact with ulcerations on the teats of cows developed cowpox, characterized by fever and by ulcerations on the hands, and that this mild illness rendered them immune from smallpox. From this observation he conceived the idea of taking pus from the sore of a milkmaid with cowpox and inoculating a young boy with it; the boy did not react to subsequent inoculations of infectious smallpox material during the ensuing months.

The first attempts to use this preventive treatment in London took place in January 1799, and it became much more common in 1800 when Woodville, after being dubious, became an advocate of the new method. In 1800, Jenner sent his vaccine to the United States. Despite a British blockade of Europe, the first vaccinations in France took place in Alsace in August 1799, using a strain from Karlsruhe. This was followed in April 1800 by inoculations in Strasbourg using a vaccine sent by Dr. Odier of Geneva, who also sent cotton fibers impregnated with the vaccine to a professor of botany in Rochefort. Vaccinations took place in Vienna in March 1799, in Gibraltar in August 1800, in Stockholm in November, and in Hanover in December. Clearly, the practice spread internationally very rapidly, even though the first tests in England by Woodville in 1799 had not been an unmitigated success—a situation that did slow down the approval of the medical community by several months. Eventually the procedure was adopted because the effectiveness of the vaccine was so remarkable, as shown by comparing its successes and failures to the practice of inoculation. The intellectual climate of the time favored acceptance of the new procedure: the socially prominent had confidence in the new medicine, they expressed their faith in progress, and they were convinced that they themselves had an essential role to play in spreading it. Moreover, the elites in France, tattered by the successive waves of the Revolution, could reunite behind a cause that was good for their image.

The real problem related to obtaining the vaccine, for without an intense vaccination campaign from arm to arm, the supply dried up rapidly. The Duke of Rochefoucault-Liancourt, who became convinced of the merits of vaccination during his visit to London, was thereafter a forceful advocate for vaccination campaigns. He started a publicity committee and had people sign up, beginning in March 1800. In May 1800 his representatives began to bring "liquid vaccine" from England. The work then began immediately, making children in institutions the repository of vaccine fluid. Over the course of a year, it was easy to arrange that several children would have pustules at any given time as a result of their recent vaccination. This was the first time that an operation of such scope was led jointly by scientists and public authorities and also the first time that society had a simple and effective preventive measure against what had been one of the deadliest epidemic diseases of the eighteenth century.

A detailed analysis was made of the results of vaccination in the area of Liège. That city's charity office had suggested in November 1800 that the regional prefect find out about the new discovery. Augustin Thouret, member of the medical committee established in Paris, informed the prefect of very encouraging results obtained with the "new inoculation." He concluded by encouraging the latter to take action. "So, citizen, these are the results that we have gathered up to this point; we think they are sufficient to ask the government to support, with every means at its disposal, the trials that are being undertaken with all appropriate precautions. Moreover, we believe that the results oblige the government to encourage all educated people to favor increased numbers of trials so that enlightened opinion concerning vaccination may spread." In the spring of 1801, the vaccination committee sent a Liège physician to be trained in Paris. Even though the prefect had his own children vaccinated, apparently in Liège things got off to a slow start. A free vaccination clinic was established, and subprefects and mayors were encouraged to set an example so the principle of vaccination might be widely disseminated. In December 1803, however, the prefect still had to admit, "It will take more time for people to get used to the name and the idea of 'the vaccine.' They will need to see the example of a greater number of enlightened men undergoing the practice, and they will need more health officers favoring the new inoculation based upon more wide-spread experience with it." And, he added, this experience had been accumulating since 1802.[20]

Numerous bulletins were sent to mayors, emphasizing that all the experiments had shown the vaccine to be completely safe and also emphasizing the possibility of "suppressing forever this destructive scourge." The prefect even appealed to the bishop of Liège to mobilize the parish priests to educate the faithful and disdained the apathy and lack of foresight of those who made up their mind to vaccinate their children only when faced with a looming epidemic. "In my opinion, I will not have completely fulfilled my duties to the people until they consider inoculation as an obligation that parents have to their children, the same as registering their births." From that time on, civic responsibility included administrative registration of the population, as well as the protection of its health.

Resistance to vaccination was seen as a manifestation of an "ignorance that is doubly reprehensible, as it imperils both the health of the individual and that of the group." Never before had there been such a clear expression of the values that justified the post-Revolutionary regime in France. Vaccination was a sure thing, a hope, a project, and a symbol, all at the same time. The simple act of getting the vaccinal inoculation required the abandonment of prejudices and involved the building of a society where knowledge and power were founded upon identical bases. Smallpox was "that scourge, which although still dreadful, will instantly stop devouring our precious future generations, once the current generation has the good sense to will it."

The system for promoting vaccination was hierarchical, with authority emanating from Paris. In each region, the prefect in charge first mobilized physicians and prominent members of society in the regional capital; he then involved the smaller cities and towns and, finally, the countryside. But the pace of vaccination could be slowed by just a few reluctant physicians or health officers or a few charity workers advising not to go along with the vaccination movement. A new boost to the effort came in 1804, when vaccination committees were established in each subprefecture for the purpose of increasing the numbers of vaccinations. These local groups were under the supervision of the central vaccination office in Paris. In the Ourthe region, for example, the vaccination committees of Liège, Malmédy, and Huy comprised nearly two hundred people, among them physicians, pharmacists, priests, leaders of charitable organizations and institutions, mayors, teachers, judges, local politicians, civil servants, and the socially prominent. The cream of the local medical community was responsible for the management of the local committees and provided an example that encouraged the growth of vaccination-related knowledge and practice as quickly as possible throughout the countryside. And since there was no central budget, vaccination committees also encouraged towns to pay vaccinators from the funds of local government charity offices. A few months after the establishment of the local committee system, community vaccinators were in place in numerous locations in the area, and other physicians and health officers were administering vaccinations for free. The death of a health officer, however, could leave several towns without someone to administer the vaccine, sometimes for months; moreover, some people still preferred their children's protection to come from a mild case of smallpox rather than from the vaccine because they thought the disease itself offered better protection.

Sweden is often considered the exemplary case in the adoption of vaccination. The first Swedish vaccinations took place in 1802, with the active support of the king. Beginning in 1803, vaccinations were required to be financed by church collections, and priests were authorized to provide the preventive measure. Vaccination was made obligatory in 1816, even though the maximum possible coverage rates were already close to being achieved.[21] In Finland, seventy-five "vaccination districts" were established, with vaccine paid for by the central government; complete coverage was achieved as quickly as in Sweden.

A Short-lived Success?

Remarkable results were achieved in France because of the enthusiasm of the upper classes, the encouragement of those in power, and the demand for care on the part of the population. The 150,000 vac-

cinations given in 1806 grew to 750,000 in 1812. In the first decade of
the 1800s, the annual number of births was about 950,000; subtract-
ing the 200,000 deaths of infants under one year of age leaves
750,000 surviving children. Vaccination rates were thus very good,
even if those receiving the vaccine were not exclusively under one
year of age. The laudably high rates of vaccination explain the rapid
decline in the annual number of deaths from smallpox to 2,000,
rather than the 50,000 to 80,000 per year just before the French Revolution. But any weakening of enthusiasm or of government support
led inexorably to a diminution in the numbers vaccinated. In France
in the 1820s, when the average annual number vaccinated fell to
350,000, the main factor in the decline in vaccination of young children was probably the decreased involvement of the government.

The power of the medical schools in late-eighteenth- and early-
nineteenth-century France was reinforced to some degree by the
sweeping change in attitudes toward the body. The new sensitivity to
garbage, cleanliness, odors, and other elements of the environment
could only encourage sanitary reformers who were looking to establish connections between illness and the surroundings. Such "sanitarians" pleaded in favor of public projects to clean up the environment, just as they agitated for better housing conditions and for a
new, uplifted morality among the poorest members of society. The
elites of the medical community made it a point of honor to be the
driving force behind what clearly was a progressive adoption of new
norms, which followed the framework proposed by Norbert Elias.[22]
But the level of commitment of the authorities of the central government proved to be the key factor in promoting smallpox vaccination,
just as it had been in instituting surveillance in the 1700s to control
epidemics of the plague. The involvement of the elites permitted
France to achieve a satisfactory rate of vaccination during the Na-
poleonic era. Initial mishaps, however, gave rise to doubt in the population. Moreover, since the policy of aggressive government support did not continue indefinitely, there was a noticeable decline in
vaccination as the population grew less communally oriented and
had less and less direct experience of smallpox epidemics.

CHOLERA

The Return of Epidemic Disease and the
Abandonment of Traditional Protective Measures

Cholera occupies a special place in the history of European epidemics. The most important reason is timing. The long absence of plague after the outbreak of 1720–22, the concepts of progress and the Enlightenment, the marked decline of mortality, and the lessening of epidemics of dysentery and typhus all contributed to a sense of optimism in the late eighteenth and early nineteenth centuries. Smallpox vaccination seemed well on the way to diminishing the importance of the last fearful epidemic disease. There were just a few occasional, isolated outbreaks of yellow fever (imported from Spanish colonies to Catalonia in the early 1820s and also occurring in Gibraltar and the French Atlantic ports), which continued to keep public health officials and medical experts busy.

The European countries seemed to have become havens from disease, protected (as contemporaries thought) by their standard of living and by their "high degree of civilization," as a French doctor wrote. The spectacular advance of cholera toward northwestern Europe during the years 1829–32 was seen as a kind of challenge for society but also for a medicine more self-confident than in the past.[1] Based on a vehement theoretical debate between proponents of contagion versus infection as the cause of cholera, European governments abandoned the traditional system of protection against epidemic disease that had been established five centuries earlier, at the time of the Black Death. Although Europe had been essentially uniform with respect to the organizations and means used to fight outbreaks since the fourteenth century, national differences now came to light. Liberal ideology could not accept impediments to the free circulation of people and goods, and England was the first to abandon the old technique of quarantines at borders in favor of stricter surveillance and inspection of the sick within English territory. Bitter tensions arose among the authorities trying to control the outbreak, the physicians, and the endangered populations. Distrust of the government became prevalent, and while governments sometimes backed down in the face of popular discontent, they generally imposed extremely strict control measures on individuals.

Contagion or Infection?

For a long time, the learned explanation of epidemics was divorced from the popular explanation. The popular view was a kind of empirical knowledge, based on long experience, involving the role of contagion and the risk that people or merchandise coming from infected areas might carry disease. The learned focused on other factors: in the medieval period the conjunction of stars was seen as important; then, with the revived influence of Hippocratic theories, various aspects of the environment took center stage in the eighteenth century. Girolamo Fracastoro (1478–1553) wrote works on the contagion of syphilis, including a remarkably insightful description of "germs," explaining that they must "have the ability to multiply and spread rapidly," even though they were never seen in his day. Nevertheless, the learned were inclined instead to believe in "epidemic constitutions," a theory dear to the neo-Hippocratic movement. They established a strict and direct correspondence among the sick person, nature, and living conditions. Infection, the cause of illness, is the result of conditions in the environment, they said. And changing the air or the surroundings is truly the way to permit and promote the cure of the patient. The neo-Hippocratic physicians thus contributed to the strength of the aerist movement. According to the aerists, contami-

Facing page. The cholera pandemic began in Bengal and reached northwest Europe through Russia, the Baltic, and central Europe. Cholera spread to the Americas via Europe.

From *Une peur bleue: Histoire du cholera en France, 1832–1854,* by Patrice Bourdelais and Jean-Yves Raulot (Paris: Payot, 1987).

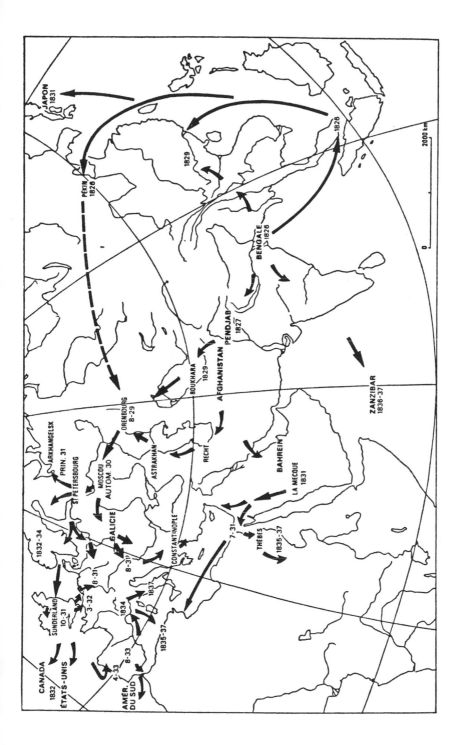

nation and illnesses arrived more often by air than by water; dust
and pollen were particularly harmful. Around 1800, the various dusts
suspended in the air were strongly suspected of being the cause of
contamination. For those who believed in the theory of contagion,
danger came in the form of people and goods moving around and
transmitting illness. For adherents of infection, only atmospheric
and local environmental conditions were responsible for new cases
of sickness.

During the 1820s, the theory of contagion seemed unscientific in
terms of the physiological medicine advocated by François-Joseph-
Victor Broussais. Broussais, an eminent professor at Val-de-Grâce,
opposed the contagionists because he believed that contagion implied the existence of a sort of natural being, an illness-demon that
took possession of the body. This went against the beliefs of his
"physiologist" school of thought.[2] The physiologists believed that all
acute illnesses arose from gastroenteritis, a condition readily identified at autopsy. With its emphasis on autopsy, the physiologist school
flourished in Paris, which was the acknowledged center in Europe for
clinical and pathological anatomy.

The physiologists believed it necessary to seek out the root cause
of the gastroenteritis causing any illness by choosing from the neo-
Hippocratic roster. It could be a matter of atmospheric changes or,
perhaps, unhealthy surroundings and the deterioration of the body, a
sort of compromise between the position of Sydenham (the "English
Hippocrates") and Villermé's investigations, which were more focused on the social environment. According to this neo-Hippocratic
view, a cluster of people who have a disease can only represent a secondary cause of an epidemic; the principal cause is the set of local
conditions. From the public health point of view, these conditions
should be neutralized and disinfected and also avoided (by leaving
for healthier places, for example). From the individual point of view,
Broussais recommends the "anti-phlogistic" method, which elimi-
nates inflammation by the use of leeches and by frequent massive
bloodletting.[3] Inflammation has a local focus and radiates outward;
the seat of inflammation produces, sustains, and propagates the infection that has caused the irritation in the body of the infected peo-

ple; they become suffused with the miasma emitted by the inflammation's source.

Pierre-Fidèle Bretonneau (1778–1862), however, changed the way people thought about contagion. His theory of morbid specificity held that a given contagious disease was the result of an attack by a particular germ. Even if a specific type of inflammation indicates a specific illness and a specific cause, it is still necessary to seek out the specific disease-causing organism. This was not a simple revival of Fracastoro's idea, for contagion was now proven, based on the meticulous observation of sequences of transmission between individuals.

There was now evidence that an infectious agent could act in two ways: direct contagion (person-to-person contact) and indirect contagion (contact through an intermediary such as another person, clothing, or the air, for example). Pathology was seen as degeneration by Corvisart and Laennec, but it was seen as generation by Bretonneau, because the illness is rooted in living beings that are a form of life even if they produce particular signs of illness. The invading germs are held by the body, which can sometimes free and spread them. Contagion is spread by means of the body, but what grows within the body and is called "contagious illness" is really the specific illness, and it is the result of contagion and not infection.

The emphasis on the importance of the contact is the consequence of a distinction made between the germ that causes the disease and the vector that spreads it. For the germ, the sick person is only one temporary support among many. It thus seems logical that, to break chains of transmission, one could isolate the carriers of germs and keep them separate, even at the hospital. This theory stands in contrast to the infection theory, where such measures were seen as conducive to the generation of foci of disease, dangerous for people living in surrounding areas. So one theory would encourage isolation of sick people to avoid contagion of those who are well, while the other theory would discourage isolation to prevent the formation of foci of infection. The two theories lead to diametrically opposed courses of action with respect to the collective struggle against disease.

Contagion not only was a revival of an old theory but also implied a revival of the old system of protections that had been put in place against the plague, involving numerous restrictions on the freedom of trade and travel. Once again, blockades and quarantines of ships and people seemed necessary. At the start of the nineteenth century, calls for scientific and political progress tended to coexist among the elites: many physicians provided simultaneous support to the Enlightenment humanism movement, to the medical revolution, and to the political and economic liberalism brought to France once more by the revolution of 1830. The sudden shift from one theory to another during 1831 was thus the result of many factors. It occurred as a cholera epidemic advanced toward Western Europe.

The Cholera Epidemic as a Natural Experiment

In March 1831, the debate on the contagiousness of cholera grew more vocal. The epidemic provided yet another occasion for adherents to and opponents of the theory of contagion to test the theory against the observed characteristics of epidemic disease. The Academy of Medicine was divided by a request for information from the sanitary bureau of Marseilles, which until then had been following the traditional practice of quarantining all boats suspected of having cholera on board, according to the old rules dating back to the time of the plague. In reply, Dr. Pierre-François Keraudren noted that the point of view had to specified in advance of a reply, so that it could be understood whether a contagionist or anticontagionist viewpoint was sought. The learned academy appointed a subcommittee of its members to formulate a reply. The uncertainty was heightened by French physicians' limited knowledge about the disease, since most knew only what they had heard in accounts and reports from foreign colleagues who had faced the illness in their countries in 1830 or early 1831.

Medical questions in France at that time tended to be addressed differently depending upon the region or town. Alexandre Moreau de Jonnès, the secretary of the central Sanitary Commission and later of the governing Health Council from 1820 to 1831, was convinced that the new disease was not an "infection" because it was only weakly affected by climatic or atmospheric conditions. Rather, it was an interpersonal "contagion" that "possesses the power to develop and reproduce under special conditions (as with more complex organisms), and which spreads either directly or through an intermediary from a sick person to someone who is well. It is thus to be termed not an epidemic, like malaria, but a contagion, like the Plague of the East."[4] This position was widely shared at the Academy of Sciences and at the Academy of Medicine until the spring of 1831. Moreover, cholera was considered a contagious illness in the first published reports from Russia concerning the disease. Dr. Reimann, director of sanitary enforcement in Saint Petersburg, went so far as to write that he "didn't know of any example of villages that had been struck without direct contact" with other locations that had been previously infected.

The opposite point of view, however, was also expressed early in 1831, as reports from Moscow began to accumulate. French physicians, sent on a scientific study mission, wrote that "the illness can be neither imported nor communicated." The Muscovite physician Dr. Jachnichen became the strongest opponent of the contagion theory and against the views of Moreau de Jonnès, crediting his own report on the outbreak of cholera in Astrakhan with having changed the thinking on this subject, convincing the government that the Asian illness was contracted in a different way. He believed that a specific "epidemic constitution" had been present in the atmosphere before the outbreak in Moscow, as shown by an overall increased incidence of diarrhea and vomiting. He also believed that miasma entered the organism through the respiratory tract, rather than by contact with the skin (which would have been contagion). It was atmospheric conditions, air currents, and the like that favored the long-distance spread of the disease as an epidemic. According to Jachnichen, the only really useful measure against cholera would be to put all patients with the disease in hospitals, thereby preventing the buildup of toxic emanations in the air of homes, which would endanger families. This supporter of infection—or rather, of penetration—apparently had no problem with creating foci of emanation in hospitals.

The distinction made between contagion and penetration ultimately led to the conclusion that quarantines, blockades, and sanitary regulations were useless (although regulation of people moving into cities or countries needed to be stepped up). The main objective of the distinction seems to have been to fight the Western European suspension of trade with Russia. Jachnichen admitted that he was very upset by the "appalling news for trade, that ships arriving in Marseilles from Odessa have to be quarantined at their destination, for a full forty days. Is this to be yet another costly consequence borne by businessmen thanks to some of the reports about cholera?"[5]

There were other examples of pressure for relief from restrictions on international trade. Early in 1832, Professor Delpech, from Montpellier, heard reports that cholera had been imported into England and Scotland. But, he wrote, any English physicians who believed in contagion were pressured into silence by the English business community, which feared the losses that such a view would cause.

Some people are not susceptible to cholera, and it is very difficult in any case to trace a chain of transmission in a big city; these factors made it difficult to reach definitive conclusions about transmission. Some physicians experimented on themselves to prove their anticontagionist views: they would lie on beds where patients had died of cholera or inject themselves with the blood of cholera victims, as did Dr. Foy in Warsaw. Of course, only those experimenters who survived their experiments were in a position to report their results to colleagues and the press. Dr. Foy "suffered only a mild indisposition" as a result of his experiment; word of this outcome was spread more widely than the report of a young Berlin physician who died after receiving the blood of someone with cholera. Brière de Boismont, spokesman for the French medical commission visiting Poland, examined the question of contagion and concluded, "The view that contact with a sick person transmits cholera to a healthy individual is false. I have touched hundreds of cholera patients; I have inhaled their breath." This was the most frequent argument made against contagion: it was believed that, if the contagion theory were correct, every contact with a patient would inevitably be followed by the appearance of the disease. Brière de Boismont adds, however, "Since cholera has been introduced by infected people in the overwhelming majority of places where it occurs, it seems natural to conclude that isolation of infected individuals (or of those suspected of contamination) is the first protective measure that should be adopted."[6] In October 1831 he proposed setting up blockades and isolation hospitals while acknowledging that they usually offered only partial protection because of imperfect compliance. Supporters of the infection hypothesis were not disposed to accept these recommendations. In its report in the summer of 1831, the Academy of Medicine was clearly divided, and when cholera struck Paris in April 1832, the issue of contagion was not yet settled.

Nevertheless, from that point on, all political and medical authorities supported the view that cholera was not a contagion transmitted from person to person. After the deaths of the head of the food service, a nun, and a nurse at the Hôtel-Dieu hospital in Paris on April 10, there was concern about not being able to attract hospital personnel; at every discussion and training session, the official view of the noncontagious nature of cholera was affirmed. It seemed at the time to be the prevalent opinion in the medical community. But with a new resurgence of the epidemic, those who had been trained by Bretonneau and favored the contagion theory became more vocal. For example, Velpeau, chief of the cholera service at La Pitié hospital, came out clearly in favor of contagion. "Out of eighty-odd cases in town whose main characteristics I had recorded, there were none without contact, whether direct or indirect, with other cholera patients."[7] This sort of observation became more and more common in the press. Doctors officially adhered to the belief that cholera was not contagious, even if they were not convinced of it on the basis of scientific evidence—they were much pressured by public authorities and the hospital bureaucracies because of concern about retaining nurses and other hospital personnel. By their silence or by their adherence to this view for political reasons, they contributed to the political management of the epidemic. But they also set themselves in opposition to the beliefs of the population at large.

Health through Isolation

Opposing medical views were being put to the test by the circumstances of the epidemic, and there were pressures from big business, but these were not the only factors in the temporary switch in the prevailing theory (which the general population didn't believe anyway). From the political point of view, it was important to stave off panic and its economic and social consequences. Nine months before the first cases in France, the better-informed segment of the medical community was on the alert, thanks to the accounts provided by Brière de Boismont beginning in July 1831. He reported that Poland was struck very hard by the epidemic, approaching the city limits of Opatow. The French commission sent to the scene as observers found that the population of the area was armed and had also prevented any travel to or from the vicinity. "It looked awful there, all business had stopped, all shops were closed, a gloomy silence prevailed everywhere; every residence had dead, dying, or sick people in it . . . The first concern of the commission was to resuscitate the courage of this overwhelmed group of people . . . They went to the authorities and the inhabitants of all the nearby places, and showed them that the illness was not contagious (from person to person) at all. Travel resumed, the markets were restocked, confidence returned, and the incidence of the illness accordingly fell in a spectacular fashion."[8]

The insistence on the noncontagious character of the disease seemed indispensable to the continuity of economic activity and social order as well as (indirectly) to the weakening of the epidemic. In April 1832, once an initial period of doubt and ridicule had passed—and the epidemic had begun to spread in the capital—the legislators in the Parlement of Paris rushed to finish up the business of their session, meeting day and night. Well-off families fled in their own carriages, while those who were not as rich joined a waiting list up to five days in advance (depending upon their economic level) for transportation out of town on mail and freight vehicles. Students returned to the provincial homes of their parents, while masons returned to their region (la Creuse), with the infected among them spreading the illness south, beyond Châteauroux. People in every social category fled the contaminated, dangerous capital region.

The flip side of the exodus was the occasional resistance in the countryside to welcoming these people, who had recently been among the sick. The subprefect in the town of Commercy, in the Meuse region, was upset by the behavior of his fellow citizens:

> It pains me to say it, but I have to: it has sometimes happened that, when these people have come back home [to take care of or bury their parents], others neglected them or fled from them, instead of visiting them to offer needed help or condolences, and instead of showing respect for them, as would be amply justified. Moreover, some citizens went so far as to request that the government take measures to ensure that people returning from regions with cholera not be allowed back into their own houses, and to prevent any people coming in from any cholera-infected region.[9]

Many accounts describe the abandonment of sick people and the difficulty of finding laborers to dig graves and bury the dead. Prefects and subprefects often had to use their authority to ensure that people arriving from an infected region weren't simply driven out by local people, who were sometimes armed. At times the acts of a person in authority seem to be contradictory to the official view. For example, on April 28, 1832, the subprefect of Saint-Quentin, in the Aisne region, came down with a fever after the arrival of a salesman from Paris, who was loaded down with secondhand clothes. Although "contagion" was not endorsed officially, the subprefect emphasized the steps he took to disinfect the environment when he recounted the story of the event to his superior in the government. "I had a doctor and a pharmacist purify all of the retailer's belongings with chlorine gas. At the same time, I invited the mayor to publicize the name of this merchant in every street, together with the nature of the danger and the indicated precautions."

Since the subprefects were in closer touch with the masses than the prefects were, sometimes they were in a position to make sure that public opinion became known to their superiors. In Verdun the subprefect reported the difficulties that arose when the Fifty-second Regiment arrived in the town, a situation exacerbated by the knowledge that the Moselle region had refused entry to the regiment because some of its members had died of cholera on the way there. As always happens in this sort of situation, some people showed an exceptional kindness in welcoming the strangers, while not forgetting to seek recompense from the government authorities. Over the course of several weeks, the expression "the blue terror" took hold, referring to the cyanotic appearance of the cholera patients: the skin took on a pallid or bluish tint because of insufficient circulation of the blood.

Cholera disrupts social and economic life. Because of the flight of the richest inhabitants, urban commerce and fine craftsmanship enter a period of stagnation. The outbreak in Paris also ruined the small businessman; the stands on the banks of the Seine remained shuttered, and the owners vanished. Unemployment was rising at the very moment when the prices of foodstuffs and of products supposedly protective against the illness were soaring. In rural villages, the fairs and markets were deserted as everyone tried to avoid infection. Opponents of inspection and quarantine condemned such measures precisely because they encouraged the population to behave in such a misguided fashion and tended to increase misery and weaken the human body as it faced the threat of epidemic disease.

Disease as Population Control

A previously rejected theory now regained acceptance. This reversal generated popular rebellion that exerted pressure on governments, particularly the authoritarian governments of eastern Europe. The draconian measures instituted along the lines of the old model of plague control created serious problems. First, the population at large generally refused to believe that cholera was a new epidemic, inclining instead toward the belief that the outbreak was a plot to reduce the numbers of the poor. During this period of severe social tension throughout Europe, mistrust of the authorities was equally marked in Russia, Prussia, England, and France. The sick were brought only irregularly to hospitals, where, according to the common belief, even the doctors were participating in the extermination of the poor. Tension was so high that some doctors were assaulted on the street. After the enforcement of rules concerning systematic hospitalization of all cholera patients, Russia saw riots, massacres, and looting, and the authorities quickly abandoned their initial policy. Similarly, in Prussia, strict rules concerning quarantine, blockades, the publication of health bulletins, the public listing of cases, and rapid burials were often disregarded.

In Paris, the police chief considered rumors of poisoning to be politically motivated: he saw in them the action of antigovernment agents: "I have been informed that some of the poor have taken it upon themselves to go to bars and butcher shops with vials and packets of poison, whether to contaminate water fountains and jugs of drinks, and to poison meats, or simply to pretend that they are doing these things." He concluded that it was a good idea to redouble surveillance of merchants selling drinks and meats. Two days later, the mob identified the first poisoners, who were accused because they were carrying a white powder. The mob grabbed them from the police and massacred them. Mob violence lasted several days and instilled fear in the establishment. Casimir Périer spoke of "savagery," and François Guizot saw "an immense font of barbarism."

Similarly, in the valley of the Garonne, far from the infected zone, poisoners were spotted in late April 1832. An anonymous letter addressed to the mayor of Casteljaloux warned that "four people have been given the job of putting a substance in the city's public water supply which would make the inhabitants sick and give the appearance of cholera. In return for this, each was to receive the sum of five francs." Yet the water remained untainted and no one got sick. Nevertheless, the rumor spread to neighboring towns; a few days later there was a report of

> a stranger near a public fountain, seen by two children from Marmande. The next day they saw in their water two strands of horse hair, dyed green. They brought them to a neighbor, who brought them to a dyer, who recognized the green coloration as verdigris; soon the whole town was alarmed. However, a local official arrived on the scene accompanied by a pharmacist, who pronounced the water completely free of any deleterious substance. Next to the fountain, the remains of an old paintbrush were found, still encrusted with oil paint made with white lead and Prussian blue. The discovery of this relatively innocuous source of the colored animal hairs laid to rest any vestiges of doubt or apprehension.[10]

Such scenes were common throughout France. People stood guard all night over wells and fountains and sometimes arrested suspected poisoners. But the situation had a decidedly political aspect, as opponents of the July revolution tried to stir up public opinion using the cholera epidemic as a pretext. On the side of those who had newly gained power, Armand Carrel denounced (in *Le National*) the "attempt to poison wine and meat," which he ascribed to supporters of the old guard. Cadet de Gassicourt, the mayor of the fourth arrondissement [district] of Paris, made similar accusations in *Le Constitutionnel*. On the other hand, the pro-monarchy right wing saw cholera as an instrument of divine anger, which had come to the capital as a punishment to expiate the sins of the revolution and the destruction of clerical property in particular. Biting political cartoons were popular, representing the accusations of the various sides.[11] Not until the death of the prime minister, Casimir Périer, and some deaths among women from the Parisian aristocracy did the reality of the epidemic become clear in all quarters.

Doctors found themselves in a paradoxical situation. In all eastern European countries, they seem to have come under attack from the general population, just as they did in England and France but probably for very different reasons. In central and eastern Europe, doctors were seen as agents of authoritarian feudal regimes, obeying the government's orders just like army officers or the nobles. In Western Europe, the reasons for animosity were more complex. At the start of the epidemic in France, for example, the republicans accused the doctors of being a tool of the rich, the nobility, and the merchant classes, working together with the priests toward the goal of poisoning the people. Thus, in Paris, one Dr. Pravaz was beaten unconscious and left for dead by a mob. He had tried to convince them that the epidemic was a reality, a natural outbreak, and that the government had nothing to do with the spread of cholera. A young medical student, assigned to a medical office in the neighborhood of Saint Germain, was stabbed and thrown in the Seine, and several other physicians were threatened.

In England and France, people were aware that cadavers were in demand for use in hospitals where anatomical pathology was taught through autopsies. Consequently, cholera was immediately seen as a means employed by the medical establishment to obtain bodies. Furthermore, patients and their friends and families feared the physicians' experimental treatments. Accordingly, mobs sometimes intercepted stretchers carrying the sick to the great Hôtel-Dieu hospital in Paris, or invaded Russian hospitals, or dragged sick patients from the hospital in Memel (a Baltic town) to return them to their own homes. Measures such as mandatory isolation and hospitalization of the sick were much less well accepted by the poor than by the well-off, who were able to evade them, providing yet another demonstration of social inequality, to the disgust of the masses.

Doctors encouraged the development of regulations that interfered with respectful mourning and funerals and sometimes mistreated the body itself by failing to show the respect due to the departed. In Danzig, all those who had died of cholera, regardless of religion, were buried together in a mass grave in an open field that was not consecrated ground. Bodies were even handled with metal tools, a violation of the practices of several religions.[12] The population's sensibilities were offended when funeral corteges were suppressed, funeral rites were shortened to a brief ceremony, and lime (calcium carbonate) was scattered as a disinfectant over bodies in mass graves. Moreover, the extreme rapidity of burial often cast doubt on the fact of death. In all countries, there were uprisings when the people had the impression that their dead were not being treated with the respect and dignity required by traditional observances.

On the other hand, once the presence of the epidemic had been announced, the population wanted and accepted help from any quarter, including nuns, physicians, and even medical students. Surviving from this period are many letters from mayors seeking support, asking regional or national authorities not to abandon them. Far from big-city hospitals, where autopsies for research and teaching purposes were common, people respected establishment medicine far less and frequently sought help from empirics and local healers.

The Mobilization of Political and Technical Resources

During the 1820s, councils of health and of public health were instituted in regional capitals, modeled on that of Paris, which had been established in 1802. But they seem to have worked only intermittently and with minimal effect on public health. Once cholera entered France, the government decided to create a public health infrastructure consisting of hierarchically structured councils of health, for *départements*, subprefects, and cantonal areas.

In the Yonne region, the three levels were put in place early in April 1832. The principal responsibility of the cantonal council was to transmit the initiatives originating at the prefecture level above them, but the council was also expected to "collect any observations that might be useful to know" and pass these along to authorities at the national level. The project of the Royal Society of Medicine, at the end of the ancien régime, had been to collect information on disease; this project gained new life in the form now set up by the administration. The centralization of observations offered the opportunity, at last, to test various hypotheses and improve the level of knowledge concerning this epidemic disease. The départemental health councils gave an important place to medical practitioners, veterinarians, pharmacists, and professors of chemistry and physics. Taken together, these scientific personnel made up more than half of the members of health councils; the rest were mostly city officials from the départemental seat.

The chief of police in Paris made an individual by the name of De Moléon responsible for publicizing the official decisions and undertakings of the Paris public health council. De Moléon required local councils to report to central authorities the "sanitary measures and general arrangements which would make the best use of the medical service, given the resources and customs of the area . . . The authorities are hoping for much progress to be made by our 338 local health councils, all dedicated to the same goal and using uniform methods."

The organizational hierarchy was rather well designed to achieve the goal of gathering information on the spread of illness, thereby improving scientific knowledge. It served equally well the purpose of creating a central administrative network for the effective dissemination of official medical and public health directives. The main thing was "to prevent isolation, which is a source of ignorance, misery, and the inability to do good," stated the prefect of Deux-Sèvres on May 4,

1832. Thus, detailed descriptions of the first cholera patients were sent to all prefect-level councils, as an aid to country doctors and rural health officers in making diagnoses. Also circulated were brochures suggesting treatments for mild and for severe cholera, together with the precautions to be observed during convalescence. One of that era's best-known works on cholera, by Dr. Bouillaud, a professor of clinical medicine in Paris, was also attached. Prefects passed this information along to the local authorities with strong recommendations for the recipients—local health officers—"to study carefully the material presented, which was produced by our greatest experts."

When the epidemic hit, physicians demanded a greater presence on the councils and in public health decision making. In addition, the power of the medical hierarchy was reinforced, to the advantage of the professors of medicine in Paris. The different councils served as a conveyor belt, bringing information from Paris to the smallest community, whose individual houses were visited by members of the cantonal councils. To avoid the total incomprehension that they would have otherwise encountered, the local authorities had to exert themselves to explain the rationale behind the public health measures that had been taken.

Turning Away from Traditional Protective Measures

In Russia and in France, the need for public health measures to be acceptable to a restive population was one of the main reasons for the official view that cholera was not contagious. The theory had certain advantages, such as being able to reject consistently all restrictive control measures, since these measures were often unacceptable to the population and caused a lot of trouble. In Russia, peasants attacked officers who were maintaining roadblocks; in Prussia, the people revolted when food prices rose due to quarantines. Moreover, it quickly became apparent that one could not effectively maintain a "cordon sanitaire" running for hundreds of miles, cutting through towns, and sometimes even separating a house from its garden. Mobilizing the troops necessary to enforce a quarantine line effectively

—sixty thousand men on the eastern border of Prussia—would involve high costs that would make the isolation tactics impractical if they extended beyond a few weeks' duration.

Less-stringent public health control measures were adopted in various places once the epidemic's "front line" had passed through and continued along on its westward track. In Provence and Languedoc in 1720–23, successive blockades had been put in place as epidemic plague advanced, but this process was not repeated now. Compared to the earlier period, there was a lack of political will for fighting the spread of infection to new areas using the most effective means: the public strongly opposed restrictive measures. One factor in the lack of political will was the lower mortality from cholera as compared to plague. The dramatic symptoms of cholera and the horrible deaths that sometimes occurred did make the disease terrifying, yet losses in all of France were 100,000 dead in 1832 and 500,000 for the entire century. Plague mortality was more striking: in Marseilles alone, 45,000 of about 100,000 people died in the plague of 1720–22. Cholera, by contrast, killed roughly 19,000 of 1 million Parisians in 1832. The scales of mortality were strikingly different.

Both eastern and western European governments rejected the strict regulations that had been used in fighting the plague. Why maintain quarantines and military blockades at borders, when they had proven ineffective or even harmful? Why have draconian rules about declaring people sick and isolating them, if contagion doesn't seem to occur? Why require rapid burial, at the risk of inciting violent reactions, when the political situation is already precarious? In all of Western Europe, the measures for fighting epidemic disease that had been inherited from the time of the plague were abandoned over the course of a few months. Unlike earlier historians, we know today that there wasn't much of a difference between the reactions in the authoritarian eastern European countries, which remained attached to rigid principles, and the more liberal western European nations, which abandoned isolation hospitals and quarantines in favor of more effective and progressive public health policies.

Russia, Prussia, and Hungary were the first countries hit by cholera and, thus, the first to experience both the epidemic and the consequent violent uprisings among the people. They were also the first to weaken their regulations and to discontinue the prosecution of those who broke them. On the other hand, measures concerned with public sanitation multiplied. For example, in Austria, there was improved control of drinking water, and requirements were promulgated concerning the cleaning out of sewers. In Prussia, daily cleaning of streets and public places was required. In Berlin, food and medicines were distributed free of charge to the poor. Thus, the policies adopted against cholera did not correspond in a simple fashion to the type of regime in a particular place. Administrative traditions did come into play, but each country also examined the results of what had been tried in neighboring countries and decided upon its own policies accordingly.

Similarly, it is important to avoid simplistic generalizations when discussing the positions of various interest groups and the population at large. It is true that merchants liked to see the end of quarantines and, indeed, of anything that tended to block trade. But they also knew that, if a ship set sail, it had to be coming from a port with a reputation for being disease-free, or it wouldn't be allowed to dock elsewhere. Rigorous and credible sanitary inspection of their own ships was necessary, for their acceptability elsewhere depended upon it. Decisions of this nature were made in the context of numerous conflicting pressures and with the hope of preserving the good reputation of the town in the eyes of foreign business partners, for a good reputation was necessary to avoid the suspension of trade. Confidence was an essential factor in international trade. The British, often protected from epidemics in the past by their island isolation, voluntarily maintained inspections and quarantines at Malta and Gibraltar.

As to the population at large, there was some belief in contagion, and the people tried to protect themselves from the new epidemic by isolating themselves. Thus, they were generally favorable to quarantines and blockades, which they believed had protected them—after all, the epidemic had not struck them. In the large towns, however, the people generally felt vulnerable to outbreaks, while at the same time they were very dependent on trade. They feared unemployment and rising prices for food. This group suffered from coercive and restrictive regulations like quarantines and blockades, and they opposed such measures. The position of prosperous merchants was probably more complex than the simple views we have been able to describe and probably depended upon the place and time in question.

By provoking the establishment of commissions and health bureaus, the cholera epidemic provided the opportunity to construct a medical bureaucracy throughout the Western European countries. The epidemic also cemented links between the physicians and the powers that be, for their statements on the nature of the illness were shaped by the political and social constraints of the time as much as by scientific convictions. Once the population as a whole—not just the elites—started to demand medical care, the epidemic propelled the physician into a dual role. On the one hand, the physician represented the hope for better health or a cure; on the other hand, he was an individual concerned with his own status, power, and income. More generally, the abandonment of the traditional constraints on travel that had been imposed during the plague years allowed the development of new methods of epidemic control.

THE "ENGLISH SYSTEM"

New Methods Gain Acceptance

In the decades after cholera arrived in Europe, a new policy for epidemic control was adopted to fill the gap left by the abandonment of the centuries-old system involving quarantines, quarantine hospitals, and blockades. A truly systematic approach arose in England, based on limited quarantine: identification of cases was followed by mandatory hospitalization of the sick to avoid spread of the disease. Several countries on the European continent also adopted this method. But how could Western European powers abandon the strict old system completely at the same time that they were trying to convince the eastern Mediterranean trading nations to enforce it ever more stringently? And, moreover, didn't the "English system" assume that the imposition of a set of stigmas was a prerequisite for successful control of disease?

The English Initiative

In England, health and social conditions had deteriorated markedly before the arrival of cholera. Many physicians, not just those who were politically radical and sometimes close to the Chartists, pointed out the negative influence of industrialization on the health of children and adults alike. In the 1830s, several commissions were appointed to study the causes of endemic fevers raging in certain neighborhoods (especially working-class areas). At a time of extreme social and political polarization, Edwin Chadwick offered a single convincing explanation for the persistence of epidemic fevers: the problem basically arose because of dirty conditions. Standing water was not drained, garbage was not taken away, and wastewater was not removed properly. Chadwick had been secretary to Jeremy Bentham, and his sanitary doctrine was a typical view for a late-eighteenth-century sanitary reformer. His main idea was to keep all the water in the city flowing freely so that wastes and trash would be cleared out. Accordingly, there was a need for massive waterworks projects to bring clean water into the cities and for sewer networks to carry away polluted water, discharging it far from the urban population.

The massive undertaking of water-related projects appealed to those in power and was given an organizational framework in 1848 with the creation of the General Board of Health. This was the first institution in Europe to hold nationwide authority for all questions concerning public health. Chadwick was its first director. After serving for six years, he was dismissed because of accumulated resentments against him and technical difficulties in building the networks of sewer pipes, as well as the inadequacies of numerous sewers (which clogged rapidly even though they had been built according to the technical specifications of Chadwick's engineers).

It is important to see this sanitary project in a much broader political context. According to Christopher Hamlin, Chadwick provided the Royal Commission with a new way out of the social and political crisis then gripping England.[1] His sanitary movement softened the shock to society that resulted from the reforms of the Poor Law in 1834. These reforms had ended all assistance to households, requiring that indigent people go to workhouses to obtain assistance. A view that was becoming widespread among physicians held that temporary unemployment led to undernutrition, which led in turn to long-term indigence, which gave rise to disease. This position was considered to be in opposition to the Poor Law reforms, which prohibited direct help to a household in need—the doctors' view implied that provision of food supplementation to families would be of value.

Chadwick offered an alternative viewpoint. In 1838, as secretary of the Poor Law Commission, he asked two of his physician friends in the Benthamite circle and one of the commissioners to investigate the causes of a year-long epidemic in London's East End, which was depleting the resources of social welfare institutions. The report pre-

pared by the three gentlemen—Neil Arnott, Southwood Smith, and
Philipp Kay—concluded that stagnant wastewater and piles of rub-
bish were to blame.

Queen Victoria reacted to this report by requesting a national
study. Chadwick was put in charge, and he sent along a copy of the
original report with every set of survey documents that was sent to
the field, which meant that respondents tended to find the antici-
pated conclusion. After all, it was clear he wanted the link between
fevers and lack of cleanliness in poor drainage areas to be confirmed.
Moreover, his overall report omitted mention of local medical conditions and of working conditions in particular industries because he
believed that local studies lacked sufficient statistical rigor and had
poor estimates of the population at risk in each location (which, of
course, would have been difficult to obtain). Finally, he systematically employed a rich variety of medical theories to ensure that each
local report focused specifically on the *sanitary* aspects of the situation and not on workplace conditions.

Chadwick cut many corners when it served his grand sanitary
schemes and also when it supported his economic and political view-
points. He had a bitter fight with William Farr, another key figure in
English statistical and sanitary history, who occupied the recently
created post of registrar general. Farr had kept "died of hunger" as an
official cause of death, but Chadwick bitterly opposed this. Chadwick questioned the relevance of such a category and gave a detailed
description of sixty-three deaths that had been attributed to hunger,
explaining why this categorization was not medically useful. Farr
replied that hunger, as a cause of death, was a meaningful description
for a large group of people because those who lack nourishment also
lack heat, warm clothing, and all other necessities of life; the hungry
focus solely on the relentless, desperate search for food, forgoing all
else. Farr's argument implied that the Poor Law Commission was a
failure! Nevertheless, Chadwick succeeded in getting public health
policy reduced solely to a concern for improved cleanliness.

Cleanliness or Poverty?

Thanks to Chadwick's reduction of public health to a single dimension—cleanliness—the General Board of Health was created. In France, the situation was very different. While Chadwick was running the General Board of Health, the Paris medical school was enjoying one of the most glorious moments in its history. The prominent professors of anatomical pathology made use of the concentration of great numbers of patients in enormous hospitals to increase their observations on the body by the practice of systematic autopsy. The search to understand the link between knowledge of the human organism and the causes of illnesses attracted numerous foreign students to their dissection amphitheaters. Meanwhile, the "numerical school" of medicine, led by Pierre Alexandre Louis, argued for statistical analysis of data concerning symptoms and the effectiveness of treatments. Theoretical and practical debates became more complex.

All of this activity occurred in the context of the industrialization of France. The industrial revolution stimulated concern about the effects of the economic and social changes in France, just as it had in England during a somewhat earlier period of industrialization. In France, these concerns were voiced by groups of reformers and (on the other extreme) by those who had never accepted the French Revolution and wanted nothing more than a return to an agrarian society organized along the lines of the ancien régime. The reformers included Catholics, who believed in relying on charity and on private enterprise to take care of the poor. They also included economists who turned to the state to regulate the new industries and the resulting social structure. Society became embroiled in a great debate over how to deal with the new circumstances. Demographic data (such as those on death rates and census figures) emerged as an important way to test hypotheses advanced by neo-Hippocratic medicine, as well as those advanced by supporters of republican ideals, such as the care of the poor. The Academy of Medicine took a great interest in the demographic statistics pertaining to population movements from 1817 to 1821, which had been presented by the statistician Frédéric Villot, and asked Villermé to produce a report on the subject.

Louis-René Villermé wanted to display data on mortality and the proportions of families living in poverty according to urban "arrondissements" [local districts], which he could then compare and rank. The difficulty lay in finding a measure that would clearly indicate relative poverty or well-being. He settled on the proportion of houses or apartments that were not taxed, out of all houses or apartments that were rented in a given district. This initial effort, which demonstrated an association between poverty and mortality, led Villermé to look for similar relationships in other populations. Based on these studies, he wrote *Mémoire sur la mortalité dans la classe aisée et dans la classe indigente,* which was published by the Academy of Medicine in Paris in 1828. He also wrote *Mémoire sur la mortalité en France.*

Villermé based the latter work on unpublished data provided to him by the government, on which he was able to carry out a detailed statistical analysis involving observations collected over geographic regions of varying scale. He compared mortality in two Parisian arrondissements (the first and the twelfth), in two streets in the ninth arrondissement, and in twenty-seven French départements [province-like regions]. He meticulously compared death rates in homes with those in hospitals. He estimated these rates to test his specific, explicit hypothesis: for each arrondissement, deaths in institutions would be proportional to the number hospitalized, and the numbers of elderly men would remain in a constant proportion relative to those who were ill. He then showed that, between 1817 and 1822, the death rate at home was 1 per 58.24 residents in the first arrondissement and 1 per 42.63 in the twelfth. As to mortality in institutions, the death rates for patients originating in those arrondissements were 1 in 42.2 and 1 in 24.21, respectively.

A possible objection to Villermé's study was that the differential might be due simply to differences in the "environmental healthfulness" of the locations, rather than to social inequalities. So Villermé's next comparison involved two Paris neighborhoods that were considered comparable on "healthfulness" but that differed in economic level. On the rich Île Saint-Louis, he observed a death rate of 1 in 46.04; in the poor Arsenal neighborhood, it was 1 in 38.36. Environmental influences were equivalent, he wrote; economic circumstances accounted for the difference. Granted, he ignored the effects of age structure; it's not, however, the accuracy of the calculations that we are concerned with here, but rather their reception at the time. Villermé also compared death rates in the rue de la Mortellerie with the four dockside streets of the Île Saint-Louis. He found inequalities in death rates wherever groups of people differed in financial resources. The differences were especially marked when one group suffered from extreme poverty. Data on mortality in the prisons in 1819 provided an additional example. Nearly all the inmates of one prison (Grand Force) had financial holdings and support from the outside; 1 in 40.88 died. Another prison, in Saint Denis, was a waystation for beggars and the homeless; 1 in 3.45 died.

Villermé tried to replicate this finding on the level of the département. He excluded those in which migration was very important and those with large, marshy, unhealthy areas. He kept thirteen "rich" and fourteen "poor" départements in his sample. Criteria for the categories were based on several measures: average annual income generated by each acre of land; average annual income per person; taxes paid and ownership of private property, per person; and distribution of income. The data eloquently illustrated the point: once again, mortality struck in inverse proportion to wealth.

Villermé was sent the mortality rates for Parisian arrondissements for 1822–26. He worked hard to get the arrondissements in line with his earlier classification for 1817–22 and published his study in the new journal, *Annales d'hygiène publique et de médecine légale*. His paper showed that none of the reasons usually given for mortality differentials among the arrondissements played an important role.

Villermé's findings flew in the face of the accepted neo-Hippocratic opinions of the medical establishment. Mortality differentials, he showed, had nothing to do with proximity to the Seine, the elevation of the land, or the composition of the soil. Neither density of housing nor population was the main factor. For him, the principal cause of mortality differences lay in the unequal distribution of wealth. His studies on mortality in various occupational groups completed his proof.[2]

Who was this Villermé? A student of Guillaume Dupuytren, Villermé served in the army of the French Empire, spending eleven years in the field on every battlefield from Spain to central Europe. He sharpened his sense of observation, adopted the experimental method, and proved to be a keen observer of the influence of living conditions on illness and death. A member of the learned medical societies of Paris, he published his memoirs on military medicine beginning in 1818 and was elected an adjunct member of the Academy of Medicine in 1823. From then on he was a visible and vocal proponent of the hygiénistes. After the cholera epidemic, which proved his point of view on mortality differences between social classes, Villermé was elected to the Academy of Moral and Political Sciences, in 1832.

Villermé was successful at spreading the belief that social factors were essential determinants of mortality differences, and because of his work this perspective even spread to political and economic leaders. Now it became clear that supplying fresh water and improving sewer networks were not sufficient to solve the problem of urban epidemics. Of course, it was more difficult to reduce inequality than to develop a grand policy of urban cleanliness. The authorities had limited resources and took only a few steps inspired by Villermé. They made some attempts to ameliorate the most dramatic declines in living standards among the poorest people living in the worst neighborhoods, where there really was a pressing danger of epidemics.

The 1850 law on unhealthful lodgings forbade landlords to rent property that was detrimental to the good health of the occupants. But resistance to this law was widespread, not only among landlords who were reluctant to make needed sanitary improvements, but also among renters, at least initially, because they sought to avoid the higher rents that would be imposed due to the costs of the improvements. Poverty leaves little room for concern about personal health.

In France, these tensions led to a series of changes of political régimes, including the revolutions of 1830 and 1848, the rise of the Second Empire, the Commune, and the start of the Third Republic. Charity took care of the most extreme cases of poverty; the social as-

pects of public health remained the focus of analyses and improvements, to the detriment of sanitation projects. The importance of
sanitary improvements in the conquest of typhoid fever was not recognized until the 1870s. Here, the supplanting of one approach by another is an exemplary illustration of the saying, "The 'better' is the
enemy of the 'good.'"

The New Quarantine

The English were not only the first to adopt modern sanitary practices but also the first to put in place a new system of quarantine.
This system, developed from 1860 to 1880, was increasingly supported by government requirements.[3] Quarantine had been present
in some form beginning in the 1840s, but boats arriving from the East
initially had to undergo just two or three days of isolation, even if
there was a suspicion that the ship posed a health risk. Starting in the
1840s, a new group of measures, part of what was known as the "English system," gradually replaced the traditional methods of quarantine. Inspections were conducted on board newly arrived ships, and
people who were potentially infectious were isolated and taken to
hospitals, as explicitly required by the Sanitary Act of 1866. Little by
little, the ports' sanitary officials were given substantial powers of intervention, including the authority to require the disinfection of
ships or the destruction of merchandise, as well as the obligatory
hospitalization of the sick.

According to the Public Health Act of 1872, everyone in England
was entitled to pure air, water, and soil. Toward this end, starting
in 1883, all passengers, whether sick or healthy, were required to dis-
close their identity and to provide the addresses where they could
be contacted during the five days after their arrival, so they could
be checked later for symptoms. These rules were designed to prevent
the importation of illness, but there was an even greater emphasis
on limiting the spread of disease. This is a clear example of a policy
driven by a desire to save lives, and it was rigorously enforced.
Special hospitals for isolating the sick, especially those with contagious diseases, were founded in the late 1860s. The poor who were
sick went there, too—their institutionalization was mandatory—but

more generally the hospitals were there for all those whose home life was judged environmentally incompatible with cure. For example, when lodgings were being disinfected, the occupants had to move to the hospital, at least temporarily. This English model was adopted by the continental European nations with varying degrees of determination and rigor.

In Germany, a similar system was put into practice thanks to the influence of Max von Pettenkofer. To control the spread of cholera germs, authorities attempted to stop fecal contamination of the soil beginning in the 1860s. Disinfection procedures were also used in public toilets. Enforcement was more rigorous in some places than in others. In Augsburg, homes of sick people were disinfected, several times over, by the authorities; keys to the houses were returned by the police after ten or twenty days, depending upon the prevailing illness. In Berlin, the poor could have their lodgings disinfected free of charge. The question of mandatory hospitalization of people harboring contagious diseases was more controversial. In Bavaria, the requirement wasn't strictly enforced after the 1854 epidemic of cholera. In Berlin, however, removal to the hospital was obligatory. As in England, there were attempts to remove healthy people from infected apartments, and in Prussia apartments of infected tenants had to be disinfected before they could be reoccupied.

The New Sanitary Frontier

The new European quarantine system had been adopted to lighten the burden of the regulations concerning ports. The system led to new ideas and sustained efforts to create a system of sanitary inspection for ships in all principal ports of the eastern Mediterranean. There was a move to transform the Mediterranean into a "safe" zone within which shipping traffic was certified as sanitary, under the aegis of the Western powers. The first international sanitary conference took place in Paris in 1851. Turkey and Egypt took part, but no accord was signed at the time. A series of such conferences ended with the fourteenth in 1938 and resulted in improved international cooperation in the area of quarantine policies at national borders, particularly where cholera was a risk.[4] A sanitary inspection office was started in Constantinople in 1838; it became part of the League of Nations in 1923. Another such office to serve all the Arab countries was established in 1881 in Alexandria under the direct control of Egypt's minister of health.

There was concern that the opening of the Suez Canal would offer a new navigable pathway for cholera, allowing its spread from the Indian Ocean to the Mediterranean. Just before the canal was to open, the fourth cholera pandemic was spreading. It was the summer of 1865, and for the first time the disease was moving from Egyptian ports to the ports of Western Europe. This movement was the consequence of the boom in steamship travel, which shortened travel times, as well as the expansion of trade between the eastern and western parts of the Mediterranean.

The cholera outbreak had a marked effect on relations between Europe and the Middle East, insofar as health-related policies were concerned. The Europeans wanted to halt the epidemic as close as possible to its starting point. Once cholera had appeared among those making pilgrimages, French envoys went so far as to ask the Ottomans to suspend all contact with the Egyptian ports and with Mecca. Mecca could then be reached only by camel caravan; some pilgrims commented that this slow means of travel would create an excellent quarantine. Starting with the cholera outbreak of 1865, quarantine was considered an indispensable defense against the dangers posed by the unhygienic conditions in Middle Eastern populations. By applying such measures, the Turks were transformed into the guardians of European public health. Perhaps quarantine imposed a burden on eastern Mediterranean nations, but Europeans considered this unimportant: commerce was less developed in the east, and the populations there traveled more slowly and were not thought to have the same notion of the value of time. So the Western governments expected the new requirements to be less annoying and less of a nuisance than they would be in Europe. When Egypt wanted in 1903 to bring its sanitary regulations in line with those in Europe, the Western powers refused, for they could not imagine that the Bedouins and the Arab peasants could follow the same public health rules as Westerners.

The imposition of public health rules on Middle Eastern nations and Turkey was, of course, a reflection of the international balance of power. At the time, the European countries were dividing up the world into huge empires, and they used their power to protect their hygienic sanctuary from epidemic disease. On several occasions it was explicitly said by British and French diplomats that health regulations and trade restrictions in Europe could be reduced only if Europeans had total control over Mediterranean shipping—the Mediterranean had to be a European sea. It followed, then, that two main entryways to the Mediterranean had to be controlled completely by Europe: one was the Strait of the Bosphorus and the Dardanelles; the other was the exit from the Red Sea together with the Suez Canal. The East was perceived as the source of two major epidemic scourges: plague and cholera. Westerners involved in public health imposed their observations, their knowledge, and their epidemiological theories on the East. The dominant position of Western medicine helped the West dominate militarily and politically, as well. At the international sanitary conferences, delegates' speeches described the populations of the eastern Mediterranean as being ignorant of the elementary rules of hygiene and cited their dirtiness as the source of their poverty, creating a caricature of Eastern subhuman masses who were dangerous to the health of Western Europeans. This portrait was used to justify the strict controls demanded by European governments.

At the conferences, the Christian West also expressed a horror and fear of the Moslem East. The size of the crowd making the pilgrimage to Mecca provoked a highly negative and worried response. From the 1860s to the end of the century, pilgrims were monitored much more closely. The island of Perim, in the Red Sea, was reserved as a surveillance and quarantine station, starting in 1866; in 1881 an isolation hospital was constructed on the island of Kamaran, where pilgrims were treated as if they were contaminated and underwent quarantine for at least ten days.

By 1885, the pathogenic agent that causes cholera had been identified, and the contagious nature of the illness could no longer be denied. Accordingly, attempts were made to prevent infected pilgrims from entering the sacred places. Multiple sanitary inspections were done on ships carrying the faithful to Mecca. The ships had to be disinfected and quarantined if any people on board were found to be sick. The regulations were tightened in 1892, after ships bearing infected pilgrims were discovered. The ships had been heading toward Suez, spending fifteen days in quarantine at El Tor. The ships were forced back up the canal and were isolated during the passage. They traveled under the surveillance of an accompanying steamboat, with guards stationed on the shore to kill any pilgrim trying to reach land. If an epidemic broke out anywhere between El Tor and Suez, the boat would have to turn back.

After the quarantine at El Tor, Egyptian pilgrims were subjected to an additional three days of isolation under medical observation. The policy for non-Egyptians was extremely simple: they were absolutely forbidden to disembark in Egypt. Thus, pilgrims benefited neither from the bacteriological revolution nor from the more relaxed European-style system of neoquarantine. Inspection of ships with more than two thousand people aboard was considered dangerous, so a starkly simple alternative was employed: the boat was placed under quarantine and observed to see whether an epidemic developed among the passengers. This technique was widely employed in the Persian Gulf area after the international sanitary conference of 1894. That conference focused on identifying measures to prevent the spread of cholera from central Asia to Europe. Travel to Mecca was banned in Romania after an outbreak of the plague in 1897 and was also forbidden by the British, who were responsible for preventing the faithful from leaving Bombay.[5]

Strategic and commercial interests caused European nations to develop divergent sanitary policies. Some said that only people of means should be allowed to make the trip to Mecca and that pilgrims should be required to show that they had the resources necessary to make a round-trip voyage. The Netherlands, Austria, and France imposed such restrictions. The British, on the other hand, had to take into account the religious sensibilities of Moslems in the Indian colonies, and they also had a maritime commercial interest in transporting the crowds of believers. The British decided that the pilgrimage was a religious observance that could not be restricted to the rich. The French wanted to establish rigid international sanitary inspections for ships in the Suez Canal, but the British resisted this —especially because Britain wanted to guarantee Egyptian autonomy, ensuring that relaxed policies conducive to British trade would prevail.

Naturally, the countries experiencing pressure from the West considered the measures imposed upon them to be unjust, a view that was to be expressed increasingly as time went on. Moslem monarchs were concerned that their subjects might suspect them of collaborating with the European powers to limit pilgrimages. Such conspiracy theories could endanger thrones. Moreover, the Turks and the Persians refused to prevent the re-entry of pilgrims returning from desert crossings. At the 1866 international sanitary conference, a Persian delegate maintained that quarantine measures would be more effective in Europe because of the existence of hierarchical administrative structures and distinct national boundaries. The Turks had long accepted quarantine policies but rejected the proposals of 1885 because the new rules seemed designed solely to protect Western Europe from the Near East. There were no measures to protect against epidemics spreading in the opposite direction, despite the fact that cholera had been brought to the Black Sea, Turkey, and the eastern Mediterranean in 1854 by British and French troops coming to fight the Crimean War.

It turned out to be impossible to force Egypt to become the quarantine hospital for Europe, just as the United States had found it impossible to force the European countries to serve as quarantine hospital annexes for the ports of New Orleans and New York. The United States had recurrent major epidemics, particularly of cholera, because of the regular arrival of ships carrying European immigrants. In response, the United States imposed its own inspection system. Quarantine of ships and isolation of the infected in hospitals were measures meant to prevent all outbreaks. In fact, however, the discovery of *Vibrio cholerae* by Koch in 1884 was the breakthrough that made disease control effective; testing for specific bacteria now became possible. Starting in the 1890s, immigrants to New York were checked at Ellis Island, an inspection station, where their health status was verified. For the first time, laboratory confirmation of even asymptomatic cholera cases could be obtained.

Social Stigmatization and Health

The strategy of public health physicians was to devise hygienic and sanitary measures that would prevent illness by requiring every individual to adopt behaviors consistent with the new medical teachings. Poor eating habits were the first target for improvement. For example, in 1765 Louis Lépecq de la Cloture, on tour in the Bessin region during an outbreak of dysentery, criticized the peasants' traditional heavy consumption of hard liquor.

One of the next obsessions of the hygienists was the healthfulness of homes, a preoccupation that led to the 1850 law on healthful lodgings. Villermé wrote an influential book, titled *Tableau de l'état physique et moral des ouvriers employés dans les manufactures* (Survey of the Physical and Moral Condition of Factory Workers). He noted that some workers were willing to pay more rent to obtain better sanitary conditions. "Workers of low moral standing sequester themselves in certain streets, even in the same houses, almost always choosing the filthiest and unhealthiest. On the other hand, the decent working people choose the opposite: other areas where their housing is more expensive, but where they only have contact with people like themselves."[6]

When the epidemic of cholera arrived in 1832, many physicians expressed their surprise and disgust at rural practices. In the region of Yonne, one wrote, "the unsanitary conditions in the houses I visited in Saint-Bris were exactly the opposite to everything the science of hygiene would recommend . . . To give you an idea of the disastrous conditions, I'd mention the heaps of fecal matter; certain streets are dumping grounds for this material. And there are piles of manure supplied by stable waste, as well."

The War on Syphilis

Morals and behavior were also in need of reform. Alcohol abuse was condemned, as were all forms of depravity, especially those leading to venereal disease. During the nearly one-sided bacteriological exchange between Europe and the Americas that led to demographic catastrophes in Central and South America, one illness crossed the Atlantic from West to East—syphilis. Probably carried back from the Americas to Italy by Spanish mercenaries, the disease became known by several names after wars between Italy and France. The French called it "Naples fever"; the Italians called it "the French disease."

The devastation caused by syphilis in Europe began in the mid-sixteenth century. By the end of the eighteenth century, six hundred patients with syphilis were hospitalized each year in Paris, of whom nearly one hundred died. The struggle against the illness peaked, however, in the nineteenth and early twentieth centuries. By then, syphilis was endemic, affecting one man of every ten, at least in the cities. It is more difficult to obtain an accurate estimate for the occurrence of this disease than it is for cholera or smallpox because of the shame associated with syphilis. Moreover, early in the illness, many patients became vectors of infection because they were unaware that they had it.[7] Since syphilis was associated with sexual activity and with extramarital sex in particular, it carried a powerful stigma.[8] Other epidemic diseases of the late 1800s were not characterized as divine retribution, as they had been in the past, but the venereal diseases retained that moral dimension, just as AIDS did more recently.

Many questions arose as European nations grappled with policies for the control of syphilis.[9] If syphilis was spread through illicit sexual activity, how was society going to organize the struggle against such an illness? In this case, the usual urban sanitation measures were not adequate. Any intervention resulting in protection could only be achieved on the individual scale, and social pressures could lead infected individuals to hide their affliction. In any case, wasn't it their own fault that they were sick? Do we encourage vice by trying to prevent or treat the consequences of people's failings?

Prostitutes were a prime focus of the struggle against the transmission of syphilis. In eighteenth-century France, Mandeville proposed establishing a system in which prostitutes would be regulated by the state. In 1791, the French Revolution abolished all laws restricting the activities of prostitutes, but two years later the ladies were required to register with the police and be inspected by doctors. During Napoleon's reign, inhabitants of boardinghouses for women had to be registered and examined regularly. Those found to be ill were hospitalized and treated. During the first half of the nineteenth century, this "French system" was adopted, with or without some alterations, by all other countries in Western Europe.

The British simply extended their existing system for controlling other illnesses, which had emphasized the isolation of the sick, to syphilis and the isolation of prostitutes. After all, wasn't prostitution a dangerous activity that should (like all other unhealthy occupations) be the subject of public health measures? Despite the overall ban on prostitution, the various European countries took different approaches, and sometimes the measures even differed from one city to the next. The British army was particularly hard hit by syphilis. The situation was so bad that, in the principal garrison towns, prostitutes were required to undergo regular medical exams and be hospitalized if they were found to be infected, just as on the Continent.

The identification of the gonococcus as the germ causing gonorrhea opened the way for diagnosis of this infection through microscopic examination; with improved diagnosis, estimates of infection among prostitutes climbed from 20 to 50 percent. In 1905, when the Wasserman test for syphilis was developed, it became clear that a quick clinical examination was not a reliable method for diagnosis. As for cholera, the reliability of the diagnostic test improved, but large-scale application of the test was far from practical. It now seemed worse than useless to continue to require periodic physical examinations, since they could provide a false sense of security. Around the end of the nineteenth century and certainly by the start of World War I, extramarital sex had become common in the urban middle classes. It was so common that the old system of inspections was undermined by the increase in occasional and clandestine prostitutes. Police vice squads kept hidden prostitution under surveillance. A few mistaken arrests occurred, some involving sisters, daughters, and spouses of powerful people; the resulting political problem put a halt to overly interventionist measures. Some people even called for the abolition of all regulation of prostitution.

Among these abolitionists were some who thought that the roots of the phenomenon ought to be dealt with first. In the forefront of this group were socialist thinkers and radical critics who considered prostitution a result of capitalism and of the dominance of the bourgeois family structure: the rule that young brides had to be virgins and the increasing age at marriage had combined to encourage young men to obtain the favors of female workers who needed to supplement their very low salaries. Some thought that perhaps revolution offered a solution to the social problem. In any event, it was believed that the solution to the social problem would solve the epidemiological problem as well.

More moderate thinkers believed that a solution would come from within the bourgeois capitalist regime, through improvement of the economic and legal status of women. Helpful policies would include giving women the right to vote and to sue for child support. All those who wanted an end to prostitution agreed that the current approach of regulating the activity was unfair to women, since infected men were subject to no controls and were free to go on with their lives. There was also general agreement that one class was being subject to regulation, since the customers were overwhelmingly from the upper and middle classes (which had a pattern of late marriages), while prostitutes and mistresses were poor. To the workers' movements, prostitution represented the exploitation of working-class girls (including family servants) by the men of the well-off classes. The abolitionist movement in France was the weakest such movement and the least dependent upon religious or moral prescriptions. Its members had an especially strong tendency to look for greater government intervention in the fight against syphilis.

Syphilis was a key factor behind concern about the decline of the French nation. As was also true with tuberculosis, the syphilis reporting system was impoverished. Yet physicians, statisticians, and hygienists continued to collect data for the sole purpose of alerting government agencies and the populace to the extent of the problem.

Estimates of the numbers of cases grew, stimulated by the rhetoric surrounding public health policies and by improved diagnostic testing. In 1925, eighty thousand deaths per year were attributed to syphilis; in 1929, it was claimed that the disease had cut short the lives of 1.5 million people during the previous decade, or 150,000 per year—the same toll that had been claimed for tuberculosis during the big TB control campaign at the start of the century. People made a fetish of figures. By 1943, they were attributing 200,000 deaths and a shortage of 300,000 births to syphilis.

Venereal diseases posed a real threat to public health, but clearly their prevalence was used by several groups as a political tool. Some accused the bourgeoisie of exploiting the young women in the lower class, while the bourgeoisie considered syphilis a broader social problem, imperiling the reproduction of the group. Some even accused the common folk of plotting to use the disease to destroy the bourgeoisie, transmitting it by house servants and working women to the sons of the finer families. The prevalence of the disease called into question the sexual morality of various social strata. Anarchists and libertarians battled traditionalist Christians, who advocated change in individual conduct.

Sweden adopted a distinctive position, probably because of the strength of Protestantism in that society. Starting with World War I, all sick persons, not just prostitutes, were subject to inspections, irrespective of sex or social class. Equal before the law, all those who were found to be sick were required to obtain medical attention. Implementation of this system impressed French observers because of the subordination of the individual to the collective good. The supremacy of the collective was implicit in this system and was also officially recognized by it.

In Britain, it was not only in the case of syphilis that the sick had to accept responsibilities on behalf of the group. The concept was carried to the extreme. Notification of the public was mandatory when someone was sick at home. People with contagious diseases could be held liable for exposing others to risk. Thus, people who took public transportation despite being sick could be fined and forced to bear the expense of disinfecting the vehicle. Required since the cholera epidemic of 1848, the mandatory reporting of all cases of disease expanded until, by the 1890s, nearly 85 percent of the population of England and Wales was covered. And, of course, children with contagious diseases had to be kept out of school.

Blaming the Victims: New Mothers

Young, inexperienced mothers attracted criticism. They were stigmatized by the medical and social establishments for taking poor care of their babies and were thought to be the cause of their babies' deaths.[10] Dr. Pierre Budin, founder of the clinic for mothers and infants at La Charité hospital, reported being struck by the number of mothers whose previous child had died. "In fact, once they'd left the hospital, they'd had nothing to guide them except the vast experience of grandmothers, concierges, and herbalists. Filled with ignorant notions, they had thus made mistake after mistake, and their children had fallen ill and died."[11]

Dr. Gaston Variot founded the Belleville dispensary in 1893, as well as another children's welfare institution later; he insisted on the importance of practical experience. "Especially in raising children, theoretical teaching is inadequate. Hands-on experience is essential in dealing with children, in weighing them, clothing them, changing them, bathing them, preparing their food, and in sterilizing their bottles and pacifiers."[12] In office visits, women should learn to raise a baby—what to do and what to avoid—he said. Visits to the dispensary and nursing the infant there would result in the mothers teaching each other. "The women undress their children together. Subconsciously each compares her own child to the others. This is enough to motivate those who had been careless about breastfeeding to improve quickly. The women don't talk to each other, but they do notice, judge, and compare. Given the embarrassment, no one would dare bring back a dirty, poorly cared-for child."[13] More generally, visits to the clinic became an occasion to correct publicly the most serious mistakes and preconceived notions, starting with the concrete example of babies with inadequate weight gain.

During the two decades before World War I, there was a marked expansion of programs aimed at reducing infant mortality. For the mother of that era, the dispensaries, clinics, and child welfare institutions became virtual "schools for mothers." Indeed, institutions with that name were created in Britain and the United States. In Britain the movement was launched by volunteers, but later city governments took on responsibilities like paying the salaries of "missionaries of health," starting, for example, in Manchester in 1890.[14] The role of the women "missionaries" was to visit all new mothers within ten days of the birth and to give prodigious amounts of advice. Very soon, however, "priority was given to families known to be ignorant and inexperienced, with the dirtiest and most dilapidated housing."[15] Thus, strong pressure was exerted on young mothers, especially from the working class, not only at the time of hospital visits but also at home.

The new methods utilized for the control of major epidemics were manifestations of the choices and priorities of the Western European nations. To create a vast trading zone free of epidemics, governments had to monitor individuals rigorously. The sick were automatically to be committed to "fever hospitals," and those with contagious diseases were required to report their illnesses to the authorities. The spread of epidemics was prevented by an extensive system of regulations, especially in Britain and Germany. The system presupposed the culpability of those who disobeyed any public health rules, whether by hiding their illness or by failing to obey the principles of morality and personal hygiene. A wide variety of groups—the poor, the dirty, the malnourished, patients with syphilis, and young mothers—were now all stigmatized. The stigmatization culminated in the denunciation of people from the eastern Mediterranean. They were seen as dirty, they supposedly didn't observe elementary standards of hygiene, and they posed an ongoing epidemiological threat because of the pilgrimages to Mecca.

5

From Miasma Theory to Departments of Health

The new methods established to control epidemics (particularly the cholera epidemic) were built on economic, social, and political rationales rather than on any foundation of established epidemiological knowledge. The contagiousness of many diseases was contested by the "aerist" theory of medicine until bacteriology offered a new approach that could identify the germs responsible for an outbreak. As far as the practical implications for controlling epidemics were concerned, however, the contrast between the two competing theoretical approaches was probably not as great as it appeared. After all, conditions that are "insalubrious" overlap with those that favor the development of pathogenic germs, so longstanding local public health policies founded on earlier theories ultimately played a key role in the decline of death rates from epidemic diseases.

Sanitary Reformers

The sanitary reform movement began to take shape during the 1820s.[1] Its adherents were heirs to a late-eighteenth-century obsession with preventive measures that would permit society to avoid illness and epidemics. Foremost among these measures was the elimination of stagnant air and water. The French Revolution had affirmed that the state, and consequently government authorities, had a responsibility to preserve the health of the citizens.[2] Accordingly, the medical care system was centralized in large government hospitals located in former convents, whose buildings had been confiscated. The government's mobilization of physicians was given greater impetus by a voluntarist Jennerian smallpox vaccination policy, at the instigation of Jean-Antoine Chaptal. The Public Health Council, established in Paris by 1802, comprised various experts in public hygiene, such as physicians, pharmacist-chemists, veterinarians, and administrators, as well as architects and engineers.

Paris attracted an enormous influx of laborers from elsewhere in France. The rapid growth of the city's population was the source of many difficulties, which the council was to try to address with the broad range of expertise gathered for this purpose. In general, the French hygienists' movement was focused (especially at its peak, in 1820–48) on the sanitary consequences of urban growth and industrialization. The city of Paris was the hygienists' main research laboratory, at least until the middle of the century. The leaders of the movement interacted as members of the council and lent their authority to it. They later worked together at the Royal Academy of Medicine (which had been re-established in 1820), at the Academy of Sciences, and finally, starting in 1832, at the Academy of Moral and Political Sciences. One of their main accomplishments was to start a journal, *Annales d'hygiène publique et de médecine légale,* which began publication in 1829. The existence of the journal was evidence of the emergence of a new field of knowledge and expertise. It remained the top journal in the field until the beginning of the Third Republic.[3] All leading researchers published their work in it. By their choice of research questions, they set the agenda for a new scientific discipline; at the same time, they were seeking credibility and recognition by using quantitative methods.

Alexandre Parent-Duchâtelet, a council member, gained an international reputation for his research on occupational diseases among such groups as dockworkers, tobacconists, and butchers and for his publications on the sewers of Paris. He was best known, however, for his two thick tomes on prostitution in Paris in the nineteenth century, the first sociomedical study at that level of detail. Edwin Chadwick considered him "the most prolific investigator of public health questions of our era, and the most intelligent." Until his death in 1836, Parent-Duchâtelet was also a tireless advocate of the professionalization of hygienists.

Louis-René Villermé, today best known for his book on the physical and moral conditions of workers in the textile industries (particularly cotton, wool, and silk manufacturing), had previously worked on health in prisons and on the effects of people's surroundings on their health. His studies on death rates became international benchmarks immediately upon publication, even if the details of his quantitative methods might be open to criticism.

Charles-Chrétien-Henri Marc was the most active of the council members. He wrote the introduction to the first volume of the *Annales d'hygiène* in 1829. His specialty was the medical care of those who had been asphyxiated or had drowned. He became the president of the Royal Academy of Medicine in 1833.

Alphonse Chevallier and Jean-Joseph d'Arcet, chemists and pharmacists, studied the effects of lead on workers' health and potential improvements in coin-manufacturing processes by the mint. They also investigated the possibility of extracting gelatin from bones discarded by butchers or cooks, for use as a nutritional supplement in hospitals. D'Arcet directed the construction of bathing and fumigation equipment at Saint-Louis Hospital and introduced gas lighting there. Another council member, Alphonse Guérard, made important contributions to the field of legal medicine and was editor-in-chief of the *Annales d'hygiène* from 1845 to 1874.

This group of scientists had an intellectual forum, places to exchange their ideas (such as councils and learned academies), and positions of power that made it possible for them to influence the decisions of policymakers. They focused mainly on typical problems of urban public health: water pollution, adulterated foods in the marketplace, the disposal of animal waste from slaughterhouses, cemeteries, problems with leaky sewers and septic tanks, street cleaning, housing with inadequate ventilation or sunlight, harmful new chemicals or working conditions in industry, and public services in general. In addition to many specific projects to improve public health, they were responsible for two laws: the 1841 child labor law and the 1850 law on unhealthy housing.[4] Although not uniformly enforced, these laws served as precedents for later complementary developments in legislation, which became stricter and more demanding as the century progressed.

The health reformers tended to favor either of two main systems of political thought. One, derived from the French Revolution and the Empire, involved the belief that the state was responsible for the health of its citizens and therefore ought to legislate and impose reforms where necessary. Parent-Duchâtelet tended toward this point of view. Others favored what was called the liberal ideology. They emphasized the role of individual responsibility and conscience rather than the intervention of the state. Liberals believed that the workers' standard of living would improve because of the development of industry and that their improved economic situation would cause improvements in moral and responsible behavior. Despite this somewhat laissez-faire attitude, liberals did not oppose government provision of the water supply, for example. Villermé was partial to the liberal ideology, although he did promote the adoption of a law regulating the working conditions of children.

Epidemics of cholera presented the hygienists with emergencies that demanded scientific solutions. The experts were expected to suggest how to stem the spread of illness. At the least, epidemics were opportunities for proponents of sanitary reform to present their opinions repeatedly and to persuade the leaders of many towns to make favorable decisions that would prevail for many years. Throughout France, improvements were made in garbage removal and water supply. In Paris after the epidemic of 1832, thirty-eight miles of new pipes were constructed, doubling the sewer network. These actions were consistent with the hygienists' theory: person-to-person transmission did not exist; the real causes of epidemics were unclean conditions and poverty, as well as certain climatic factors.[5]

Maternity Wards: Should They Be Closed Down?

The hygienists' view could not possibly help to explain the fierce epidemic of puerperal fever that developed in the maternity wards of Paris hospitals in the 1850s. So many young mothers were killed by the disease that some suggested closing the in-house maternity units and moving them to the surrounding countryside.[6] Stéphane Tarnier examined this epidemic from the hygienists' point of view—that is, he collected and tabulated data on specified variables using a standardized questionnaire or data form to determine whether the epidemic spread like cholera. But he proved that puerperal fever was independent of weather or atmospheric conditions and that it could break out during any month of the year irrespective of temperature or humidity. The Louis-Philippe Hospital (renamed Lariboisière Hospital by then) had been on the drawing board since the 1840s; it was specially designed so that all patients would be in small, shared rooms with a substantial quantity of air that would be periodically refreshed.

The hospital opened in 1854. During its very first year of operation, the obstetrical delivery unit had to be closed to quash a violent epidemic of puerperal fever. From this experience the question arose as to whether puerperal fever was contagious, from person to person, like smallpox. To answer this question, Tarnier compared maternal mortality in the twelfth arrondissement of Paris with the rate in Paris hospitals. The rate for mothers in hospitals was seventeen times higher than the rate for those who gave birth at home.

While remaining an adherent of the miasma theory in general, Tarnier concluded in 1857 that puerperal fever was contagious.[7] He came to believe that patient rooms ought to be isolated from one another and began to put this concept into practice starting in 1864. At the Paris Maternity Hospital, the results were spectacular. Maternal mortality was as high as 12 percent in 1870–71; it fell to 3.8 percent in 1873 and to 1.6 percent in 1875. Tarnier assumed the post of chief obstetrician at that hospital in 1867, and he opened a new building in 1876, constructed according to the strictest principles of isolation. Even before his specially designed building was available and before the exact source of the infection was known, his isolation theory had yielded dramatic results.

In 1879, Pasteur discovered that *Streptococcus* was responsible for puerperal fever. It soon became possible to test products that would optimally destroy this bacterium in different situations. Tarnier devoted an 839-page tome to such antibacterial agents, entitled *De l'asepsie et de l'antisepsie en obstétrique*.

The French hygienists' movement thus demonstrated the effectiveness of their theory, which was able to avert hundreds of deaths before the discovery of the specific germ responsible for puerperal fever. Across the English Channel, Joseph Lister published his article, "On the Principle of Antisepsis in the Practice of Medicine," in 1867, based upon a strictly medical approach and on Pasteur's initial observations. A surgeon in Edinburgh, Lister practiced preoperative disinfection starting in 1870. Pasteur's principal discoveries in the 1870s resulted in an entirely new armamentarium for hygienists to use in their struggle against epidemic disease. *Staphylococcus* was identified in 1878, a year before *Streptococcus*. In Germany, Robert Koch was developing more microbiological techniques, such as stains and culture media that were specific to different kinds of bacteria. His methods enabled him to discover the tuberculosis bacillus in 1882 and the causal agent of cholera in 1884. In less than a decade, the germs responsible for all of the main infectious diseases had been identified. The debate on the contagious nature of epidemics was over.

In 1887, the Faculty of Medicine in Paris appointed a new dean, Paul Brouardel, who worked with André Chantemesse on typhoid fever. By this time, Pasteur and his discoveries had been totally accepted by physicians. With prominent hygienists filling various public health positions, the administration of the Interior Ministry's new Bureau of Public Health and Social Welfare (founded in 1889) was well aware of the value and practical consequences of the new discoveries in bacteriology. The Society of Public Health Medicine and Professional Hygiene was created in June 1877, with a full roster of founding members who helped guide the new field of bacteriology: Brouardel, Vallin, Adolphe Pinard, Pierre Budin, Vincent Cornil, Henri Napias, Laveran, Martin, Adrien Proust, Bourneville, and Octavius de Mesnil. The new society was so successful that twelve years later it subdivided its leadership into seven permanent com-

Facing page. Hospital plan, by Jacques Tenon. After the fire that destroyed the old hospital in 1772, the new hygienists emphasized the deplorable condition of hospitals. In 1788, Tenon, a medical doctor himself, published an important proposal for a new plan for the hospital that would remedy the unhealthy ventilation and improve the circulation of air.

From *Mémoires sur les hôpitaux de Paris*, by Jacques Tenon (Paris: Imprimés par Ordre du Roi, 1788).

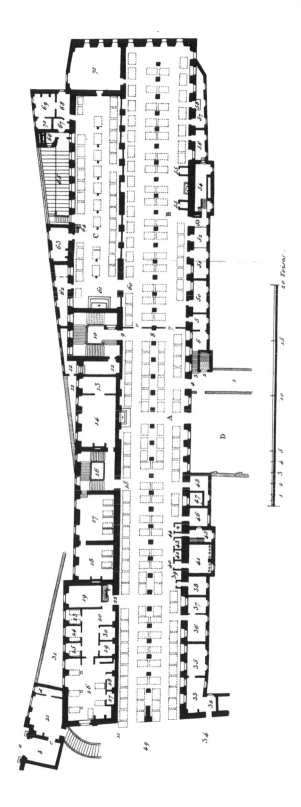

20 Toises.

15   10   5   1 2 3 4 5

mittees, each with outstanding leadership: child health (led by Henri Monod), urban/rural hygiene issues (Adrien Carnot), industrial and occupational hygiene (Henri Laveran), the prevention of contagious disease (Vincent Cornil), food hygiene (Léopold Trasbot), international and administrative hygiene issues (Adrien Proust), and demography and statistics (Émile Cheysson).

The members of the new society served together on most of the important government councils. They made up 25 percent of the members of the Paris commission on housing improvements and 40 percent of the Board of Public Hygiene and Sanitation for the Seine region. They also made up more than 25 percent of the Advisory Committee on Public Hygiene, appointed by the French national government. The society's means of promulgating the hygienists' point of view extended well beyond their journal, *Revue d'hygiène;* they had many contacts among the makers of public opinion and in the political world.

One could no longer imagine a public health movement that ignored bacteriology.[8] The new medicine was powerful. Its proponents held key positions in the relevant professional networks and had access to the press. They served on government agencies convened by the Parlement. The hygienists had a considerable influence on the administrative structure of public health in France, and they must surely be given credit for the suppression of the major epidemic diseases.

The Effects of Better Nutrition

Thomas McKeown, professor of social medicine at the University of Birmingham, set out to determine whether medicine or public health measures were important contributors to the lowering of mortality rates and the decline of the major infectious diseases.[9] It was iconoclastic even to suggest such an analysis—after all, it calls into question the heroic version of the history of epidemics, a version built on the image of great scientists making medical breakthroughs. George Rosen's *History of Public Health* is an example of this view. Originally published in 1958, it has been republished numerous times.

McKeown's point of view, expressed in a series of papers between

1955 and 1975, was quite the opposite of Rosen's. He attempted to demonstrate that medical discoveries had made no contribution to the control of endemic and epidemic diseases in the nineteenth and early twentieth centuries. Public health measures, he believed, had done little more. He stood in opposition to most other authors, who at least acknowledged the birth of bacteriology as a major turning point in recent human history. Building his argument with global figures, he easily showed that the hospital offered no advantages over home care before the changes made during the final third of the nineteenth century, in accordance with the emphasis on asepsis and antisepsis in the traditional chronology.

McKeown followed this analysis with another that was focused on specific major infectious diseases, using data on cause-specific death rates for England and Wales. His data showed that the contagious illnesses, without exception, had all declined drastically before the discovery of effective treatments or specific vaccinations. The curves were particularly striking for tuberculosis, which had been in decline constantly since recordkeeping had begun in 1838 and well in advance of antibiotic therapy, which became available after 1945. Rates of bronchitis increased until 1901 and then declined very rapidly in the absence of modern medications. Trends in measles rates were even more remarkable, showing a sudden fall after World War I, while the vaccination wasn't available until the 1970s. Only trends in diphtheria offer evidence of response to treatment, which initially was offered by Roux and his colleagues at the Pasteur Institute in 1894 in the form of a serum. Diphtheria rates collapsed under the onslaught of the British national immunization campaign undertaken during World War II.

With these findings in hand, McKeown set off to look for other factors that would account for such a broad decline in nearly all the infectious diseases in Europe. After rejecting several hypotheses, he accepted improved diet as the explanation of the improved health conditions. Better diet improves a population's general resistance to disease. McKeown pointed out that only water- or food-borne illnesses could have been prevented by the public health and sanitation policies developed in the late nineteenth and early twentieth centuries. Yet respiratory illnesses were responsible for the majority of deaths and could not have been affected by these policies, he said. This observation supports the hypothesis of improved resistance to infection, a change that could only have been due to improved nutrition. McKeown's theory is widely accepted today among physicians, who observe the strong influence of nutritional status on resistance to infection in poor populations.

For more than a decade, however, objections of various kinds to this theory have been raised, prompting a re-evaluation of the influence of local sanitary policies. The criticisms have justifiably centered on the gross scale of the data that McKeown chose for analysis. He should have looked at changes in rural versus urban areas, and he should have broken down the mortality data by sex and age. Moreover, the five main categories of "cause of death" used by McKeown were only a third of the deaths in 1851–60 and only a sixth of those in 1891–1900.[10] Among the causes of death he ignores because of their supposed "stability," some actually increased and others decreased. The "stability" is just an artifact of their being grouped together. A researcher carefully reanalyzing the data using other categories of cause of death could draw conclusions diametrically opposed to McKeown's. Scarlet fever and diphtheria were not distinguished as separate diagnoses before 1855, and neither were typhus and typhoid fever. If deaths are regrouped with a new category including all the diseases linked to the environment, the total number of these deaths is not diminished but rather remains at the same level as the number due to respiratory illnesses.

Among deaths due to respiratory illnesses, the decline over time in tuberculosis deaths was accompanied by an *increase* in deaths attributed to "lung diseases," which suggests that a partial explanation for the decline is simply a shift in the reporting of deaths from one category to another. If this explanation is correct, then changes in tuberculosis mortality would not really be the strongest factor in declining mortality: declines in tuberculosis would explain 35 percent rather than 47 percent of improvement in life expectancy at birth. The decline of other diseases—those controlled by improved sanitation—would then be the most important factor in overall lower death rates. Thus, the improving death rates, particularly in the cities, could certainly have resulted from sanitary measures and the provision of clean water, as McKeown acknowledged in his later work. These sanitary improvements could also have indirectly contributed to increased resistance to tuberculosis because children living in improved conditions have fewer infectious diseases of all kinds; they consequently tend to have better resistance because they are not chronically debilitated. Similarly, smallpox vaccination reduced the number of generally debilitated young people, and improved resistance to tuberculosis may have followed, owing to the generally improved health of the population.

Vaccination (at least for smallpox), combined with municipal investments in urban cleanliness and public health, forced epidemics into retreat between 1850 and 1940. There were a few setbacks along the way, as some decades witnessed a temporary increase in mortality in cities that were undergoing rapid growth. Public health measures to counteract the effects of overcrowding, such as congestion and promiscuity, were not immediately put in place. In the French town of Creusot, the population was twenty-seven hundred in 1836 and twenty-three thousand just thirty years later. At the start of the period, life expectancy at birth was 37 years for men and 42 years for women, but by 1856 it had fallen to 29.7 and 32.5, respectively. By 1876, the levels were back to those of the uncrowded earlier period. In this case, urban growth and industrialization had set back epidemiological and public health progress by forty years. The recovery evident from about 1861 onward occurred because the population had been made aware of health issues by the hospital and by visits to physicians and the provision of free medicines. If we add two things to the list—improvement of the water supply and required school attendance—we complete the roster of the collective efforts that reversed the negative health consequences of rapid urban growth in just twenty-five years.[11] In British cities, too, mortality improvements stalled for similar periods of time between 1820 and 1870 and then resumed.

Simon Szreter shrewdly noted the correlation between mortality improvements in European society and continuing improvements in the population's real income. So, did McKeown underestimate the influence of municipal policies that improved the cleanliness of cities? Did he neglect the value of the hygienists' policies to advance personal cleanliness? All of these policies may have contributed to the disappearance of typhus, for example, and to the decline of all the diseases spread by dirty hands. Szreter came to believe that the work of the public health pioneers during the last quarter of the nineteenth century caused the declines in cholera, typhoid, and smallpox, while increases in bronchitis may have been due to smoke generated by industrial sources and by the coal fires heating homes.[12] Definitive, sophisticated quantitative analyses are clearly beyond the scope of what can be achieved with the inadequate data from the period in question.

Despite the lack of quantitative analysis, one cannot fail to be impressed by the chronological concordance: cities adopted public health policies, and then infectious diseases declined. Many cities undertook their public sanitation projects at the same time, in the space of a few years. Of course, the sanitary reform movement did come up against technical difficulties with water supply networks and the design of sewers, but ultimately it had a positive effect on the health of the population. The movement's public works became *the* model of successful public health intervention. The French example makes it very clear: in the three largest cities (Paris, Lyon, and Marseilles), the fall of mortality in the second half of the nineteenth century coincided very closely with government-sponsored improvements of the sewers and the water supply. In England and Wales, expenses for equipment related to sanitary improvements increased sharply in the big cities from 1888 through the start of World War I. A decline in mortality at all ages accelerated in tandem with these increases. Another recent study involved records of medium-sized cities in the United States from 1888 to 1929. Every 1 percent increase in annual municipal expenses for sanitary improvements led to a decrease of nearly 3 percent in mortality.[13]

Of course, many people in political or business authority had no choice but to support the major public works projects. This does not diminish the importance of the projects in the disappearance of typhoid and various gastrointestinal ailments from the large cities. Information about personal hygiene spread gradually, as did new attitudes toward the care of the body.

City Health Departments, 1879–1900

In addition to public works projects, many of the larger cities in Britain, Belgium, Italy, and later France established urban health departments. Their primary goal was to reduce mortality, particularly from epidemics. Their tools were epidemiological surveillance, vaccination, and improvements to unhealthy housing and the water supply, both public and private.

The health departments' objectives and ambitions were articulated in the documents establishing them as official government agencies; examples include the charters from Le Havre, Nancy, and Grenoble. The motivations are also manifest in the projects suggested by the departments to city governments and in letters from one Dr. Gibert, founder of the first city health department in France, which was created in 1879 in Le Havre by order of Jules Siegfried.

Many Italian cities had had health officials since the fifteenth century; they were organized into health departments in the 1850s—as in Turin, for example. British cities were setting up health departments, too; Glasgow is a well-known example. But it happened to be the Brussels city health department, founded in 1863, which inspired Dr. Gibert. He had been sent to examine that exemplary department, and he wrote five letters about his impressions that appeared in a daily newspaper in Le Havre. No doubt this was part of a public relations campaign designed to convince the local elites and the public of the value of the project, which was an initiative launched by a new city government. Elected in 1877, the new officials presented a plan for a health department early in 1878 and appointed a commission to develop it further. Dr. Gibert went to Brussels in June, and the final vote by the city council took place on February 12, 1879. The health department came into being the following month. The letters use medical claims to further a political goal and present a stirring plea for progress in public health.[14]

In his letters, Gibert first expresses his overall admiration for the

Brussels health department. "Honestly, it would be difficult to find any gaps, or even anything to criticize," he writes. Then he shares with his readers his experience of visiting the department's director, Dr. Janssens.

> In his office, there is a map of the city of Brussels, dotted with numerous pins in many colors. The colors of the heads of the pins indicate various diseases. Each day, Dr. Janssens puts in new pins—blue for smallpox cases, red for typhoid fever, etc. He doesn't push the new ones all the way in; they are left sticking out. The map is brought to the burgomaster each evening as a sort of report on the new cases of contagious illnesses; the mayor pushes the pins all the way in.

The mayor is thus informed of trends in dangerous diseases in the city on a daily basis and can understand the situation at a glance.

Data for the maps were supplied by the city's physicians, who sent a "public health notice" to the health department when faced with any case of "smallpox, scarlet fever, measles, typhoid fever, typhus, cholera, diphtheria, or epidemic dysentery." The notice included the name, age, and address of the patient. The specific illness was entered as a code number rather than in words, as a means of preserving patients' privacy; there were 116 codes for causes of death. Hospitals, clinics, and charitable institutions also supplied information on cases. Thus, each day the city government was informed of the level and location of all new epidemics, permitting measures to be taken in a timely fashion. With smallpox, for example, everyone living in the same house and on the same street as the initial case was vaccinated or revaccinated. All the clothing and belongings in the sick person's home were disinfected, and the patient was isolated. Isolation was imposed at home as much as possible; for total isolation the patient might be sent to the infectious diseases unit of the hospital. The health department had a vaccination service, open every day and free of charge; moreover, every schoolchild was revaccinated. In the absence of a vaccine for typhoid fever, the procedure for that disease had to be different: once a case was reported, the health department immediately conducted a rigorous inspection of the water supply, sewers, and garbage collection areas in the affected neighborhood.

In Brussels, the officials regulating the quality of housing were part of the department of health and had "complete authority to force building owners to make whatever improvements the department of health may require." The department kept the file of complaints and decided whether the changes made by building owners were adequate. Gibert also emphasizes the value of information sheets filled out for each street and argues for rigorous and thoughtful collection of information on causes of death, to be verified by physicians working for the health department. Accurate statistics on the various infectious conditions striking the city could only be obtained by this method, according to Gibert.

Gibert also reports enthusiastically on the cooperation of schools with the public health system. Each school had a weekly visit from a doctor, who would identify any children afflicted by contagious illnesses of the eyes, skin, or scalp and order them kept out of school until the condition resolved. There was an educational component as well: during the visit, the doctor also gave a short talk on the illness detected or on some other contagious disease. "The pupils absorb these teachings very well, and repeat the hygiene lessons at home, and the information thus reaches their families more readily than if it were spread by books or newspapers," Gibert writes. Dissemination of health information to the public, one of the key goals of the hygienists since the beginning of the century, was now accomplished by children and the teaching they received at school.

Gibert's last letter closes with a discussion of budgetary issues, to which the city's elected officials were naturally very attentive. "At very modest cost, the government has saved many human lives and has earned the right to claim proudly that it has set an example for the rest of Europe to follow." At the same time, he reports on the salaries of the doctors employed by the health department in Brussels and insists on the importance of paying them properly in Le Havre. "Pay them well, otherwise you will get nothing. If you don't want to pay them well, it would be better not to have them at all," Janssens had told him at the end of his visit. Gibert's discussion makes his own point of view quite clear, and his comments cater to the increasingly demanding medical associations. "People will say it's expensive. So what? Is it fair for the public sector to rely on the devotion of physicians to their calling, all the time? Hospitals don't pay doctors. Is that fair? Charities pay doctors almost nothing. Is that fair? If you want doctors to work hard in the service of public health, you have to be able to require them to do so, and so you have to pay them." Such talk impressed the private doctors, especially as Gibert had created the first children's dispensary in all of France at his own expense (in Le Havre in 1875), which offered free office hours every morning. Gibert also shared the belief of public health doctors that, if prevention policies were successful, substantial savings would accrue to hospitals and free medical clinics.

In his plan for Le Havre's new health department, Gibert is arguing for medicine to have a key role in how the city is run. He is the perfect spokesman for the ambitions of the medical and public health movements of the era. "If society has not yet benefited from all the progress made by medicine in our century, the main reason must be that physicians have been denied their proper role in society until recently. It should be up to physicians to spread the news of scientific discoveries, which only they can understand and turn into practical improvements."[15]

Gibert believed that four gaps in contemporary public health services were not being met systematically by the government; these four would be the main functions of the health department. Surveillance of epidemic diseases in an ongoing fashion was the first (cholera, smallpox, diphtheria, periodic fevers, and puerperal fever are mentioned). The second was the oversight of garbage disposal (which had to be prompt and sanitary) and sewers and drains (which had to be effective and watertight). Unhealthful housing was another concern—laws were in place but not enforced, and Gibert wanted to implement the kind of effective enforcement he had seen in Brussels. Last, the health department should be involved in providing some kind of medical help, including the formation of first aid squads, who would bandage wounds and provide certain medical treatments in patients' homes, avoiding unnecessary hospitalizations. In addition to these four main functions, Gibert also wanted the health department to be involved in promoting the health of young children. He was as interested in schoolchildren having ergonomically correct furniture in their classrooms as he was in checking them for infectious diseases, and he also wanted the health department to organize public baths for children.

Gibert ends his proposals by insisting that such matters have to be handled by a health department on a citywide scale. "City governments generally do not listen to the advice of councils of hygiene, of neighborhoods, of départements, and almost never take them seriously, except for advice from their own municipal hygiene office," he wrote. He closes by reminding the reader of the importance of the proposed health department in the context of the demographic future of France. "We will feel justly proud of giving public health its rightful place in society once we see the added number of hale and hearty defenders of France, the additional workers for every industry, and the increased numbers of young women who have been well prepared for motherhood by their salutary education. We are not in first place in terms of birth rate, so we should be first in terms of the survival of our incomparable human treasure."

Pleas for the establishment of health departments in such cities as Nancy, Reims, and Grenoble were also motivated by demographic concerns, especially reports of high mortality accompanying urban growth at the end of the Second Empire. One Dr. Berlioz suggested in the late 1880s that a health department be established in Grenoble, insisting that mortality was high there compared to similar cities.[16] Berlioz noted that mortality had fallen substantially where health departments had been established, and he cited Glasgow as an example. He concluded that the establishment of a health department focused on outbreaks of infectious disease would have a significant effect on mortality in Grenoble, because fully 20 percent of deaths in Grenoble were due to contagion.

Once health departments were established, one of the first things they did was to redouble systematic smallpox revaccination. Another was to study the epidemiological characteristics of diseases using cartographic techniques because treatment and control of many diseases were not yet possible. Detailed records of where diseases struck were maintained for research purposes in Le Havre (along the model of Brussels) and in other cities, such as Bordeaux, Nantes, and Nancy Saint-Étienne. In Le Havre, statistics were published regularly on deaths from tuberculosis, typhoid, diarrheal gastroenteritis, and other diseases. Data were broken down to show incidence by season and by streets, allowing ready identification of the neighborhoods where the population was most affected by each ailment. The areas in direct need of sanitary improvement could thus be identified.

In Paris, too, detailed records of incidence were kept, identifying "killer houses," those in which mortality from tuberculosis exceeded a threshold. If several killer houses were located close together, the area was declared an unhealthy zone and the buildings were condemned to destruction.[17] It's almost as if the building was considered the carrier of the disease, in what could practically be interpreted as a return to the "aerist" doctrine! The new bacteriology had provided knowledge of Koch's bacillus, but this knowledge did not eliminate the importance of environmental conditions in the development of germs—quite the contrary. It was now clearly demonstrated that the absence of sunlight favored the development of the microorganism.

Bureaus of health did more than isolate patients with contagious diseases and keep infected children out of school. They promoted the development of municipal chemical and bacteriological laboratories, which could inspect the quality of foodstuffs, as was done in Nancy and Grenoble. They also responded to the needs of public health officials when the public was concerned about foods that were rotten or otherwise unfit for consumption. Once urban health departments became established, those cities also tended to have dispensaries, clinics, and a more fully developed health network than they had had previously.

In most cases, the health departments tried to educate the population concerning behaviors with an effect on health. In Le Havre, Dr. Gibert emphasized that "health departments should periodically draw up a table of mortality statistics according to neighborhoods, because this will make our beliefs more convincing to the public. It would be an important step toward getting public opinion in favor of the public health laws, which are all too often considered abstract and purely theoretical. Such tables should be widely publicized." Numbers, graphs, and maps were considered particularly powerful tools when it came to convincing the public.

In the establishment of a health department, reform of behavior and morality was not an explicit priority but indirectly could be obtained through the teaching of schoolchildren. Once a department had been established, however, this concern quickly took the form of little pamphlets on hygiene aimed at individuals and mothers with young children. Pamphlets were a new medium, both in format and in mode of distribution, and their delivery into households probably had some influence on the adoption of new behaviors.

One such pamphlet is twenty-four pages long, published by the health department of Le Havre in August 1880 and titled, "Disinfection Related to Epidemic and Contagious Diseases." The city council had "decided that the most common disinfectants will be available at all police stations and will be given free of charge to any indigent family upon the recommendation of the doctor treating that family." The first section of the booklet is devoted entirely to "Fresh Air and Cleanliness," which were considered "the two most effective disinfectants," so it is clear that the improvement of the practices of the poor was a concurrent goal. The next part of the booklet lists various disinfectant products and explains how to use them when cleaning the room of a sick person. There is discussion of what to do during and after the illness and how to disinfect bedding, clothing, and the room itself. The problems of how to handle the body of someone who has died of a contagious illness and how to place it in a coffin with disinfectants are also addressed.

The last part of the booklet concerns "disinfection by heat," which had been proven effective in Paris as well as overseas. In June 1880, the Seine region's health department proposed the construction of two public sterilizing ovens. According to the booklet, the hospital of Le Havre had such a facility under construction in August 1880 and the number of such facilities for the public was expected to increase in the near future.

Why did the authorities choose this type of activity as the focus for their public health campaigns? Urban political and social elites were already on the side of public health policies, so perhaps it was time to work on convincing the population at large about the importance of public health. This type of educational program also seems to provide guidance for police and other government workers in the field of public health.

Another leaflet is shorter, at just four pages; it, too, was published in 1880 by the authority of the health department. It was more narrowly aimed at young mothers. Titled "Precautions to Take to Avoid Blindness in Infants Shortly after Birth," it calls for mothers to "keep the eyes very clean." The leaflet warns mothers not to let babies catch colds and to call a doctor if any pus came from the eyes. At the same time, in the same towns, children's dispensaries were working to reduce childhood mortality, and maternal education was one of their tools. Health information brochures were most likely used for this purpose.

The Importance of Municipal Policies

Simon Szreter rightly pointed out the range of effective municipal policies governing urban cleanliness and public health. Many of these important reforms came about in the absence of legislation at the national level, and his remarks do not apply only to England. National public health laws were passed in France in 1902 after twenty years of parliamentary debates, but the big cities were proud of the measures they had already taken and conscious of their pioneering roles; they sought to keep their current measures in place. Cities that had not already taken the initiative and founded their own health departments were now required to do so if their population was greater than twenty thousand. This was a difficult goal, but the "have-nots" had admired and tried to emulate the "haves" for the last two decades of the nineteenth century. In France and neighboring countries, municipalities had regularly exchanged information on the best ways to establish citywide sewer systems, health department organizations, twenty-four-hour medical emergency units, and other desired services.

There was a tremendous demand for new services on the part of the general population. Dispensaries were overwhelmed, and requests for disinfection of homes came not only from doctors but even more from the families and neighbors of the sick. Pressure for disinfection services became so intense that the number of disinfected homes grew to three times the number of reported cases of infectious disease. Fumigation had long been practiced, but now new materials (like phenol) and new equipment (like sterilizing ovens, sometimes portable) were being used. These innovations began in 1884 when cholera was the current danger, and thereafter modern fumigation became a permanent service in the big cities. The process could be intrusive: records show, for example, that the middle class in Lyon dreaded the damage that disinfection caused to their furniture, decorations, and belongings. The innovation of waterproof bags helped overcome this problem, as did unmarked vans that brought objects to sterilizing ovens in fixed indoor locations elsewhere. Resistance to disinfection then came from those whose reputation or economic well-being could be harmed by the process—hotel owners, for example.

Basically, the authorities were surprised by the very strong demand. "Disinfection is a reassuring, even magical act, which requires no other actions and does not impose any changes in lifestyle."[18] From the hygienists' viewpoint, disinfection unfortunately did nothing to encourage people to change their behaviors, even if it did limit the spread of disease. Nevertheless, in many areas it was disinfection, sanctified by the public health law of 1902, that received most of the credit for improvements in public health.

The population was relatively easy to mobilize against an epidemic of measles, scarlet fever, or anything that struck suddenly; more difficult was tracking down or reporting a case of tuberculosis. And the population continued to reject mandatory isolation and hospitalization, with good reason; in many cases, when an apartment was disinfected, the tenant was sent away by the landlord, with housing in the neighborhood unavailable upon their return. It was fear of tuberculosis and of those bearing the disease that caused this reaction, not an aversion to the disinfection process, which we know because of the frequent requests for the process itself. The problem was that being labeled an "infected" person created difficulties in the neighborhood. To try to reduce the problem, the authorities routinely disinfected entire apartment buildings rather than focusing on specific apartments where disease was present.

Complaints about substandard lodgings were increasing at the turn of the century and becoming more and more explicitly related to medical concerns. It is clear that some progress was being made in promoting the public understanding of illness and how to avoid it, although it's uncertain whether most people had really accepted the new medical theories or whether it had just become easier to make complaints.[19] Still, in the 1890s, people made many complaints in order to get their homes hooked up to central sewer lines.

In short, the nineteenth-century hygienist movement had considerable influence on both the research agenda and the legislative agenda (at the municipal level in particular). City health departments that had been created by the big cities on their own (prior to the 1902 law) were the outstanding centers for experimentation, acculturation, and progress. Because of their consistent record of achievement and the many studies they produced, the older health departments quickly gained the recognition, and often the trust, of municipal councils and other city agencies. Their influence on local sanitary policies grew to such an extent that, even today, these pioneering cities continue to leave their distinct imprint on European urban public health.

VACCINATION

A Powerful Paradigm

Vaccination played the pre-eminent role in epidemic control policies starting in the nineteenth century, and the technique was used for more than smallpox control. No history of the struggle against epidemic disease would be complete without a chapter on vaccination. Vaccination has contributed to the control of epidemics in the developed countries from its earliest invention until the present day and has served as an idealized framework for successive victories over all infectious diseases.

Smallpox Vaccination: The Difficult Road to Acceptance

During the 1820s, the proportion of young children who were vaccinated declined, probably because health authorities were becoming less actively involved in vaccination programs. The vaccination process—involving the transmission of material from one person's arm to the next—provoked anxiety. People were afraid that the "bad blood" of orphaned babies, who constituted the reservoir of the vaccine, would be transmitted during the procedure. Some authorities claimed that other illnesses could not be transmitted via the pus used for vaccination, but there were numerous reports of syphilis being spread by this material. The controversy led to the abandonment of arm-to-arm vaccination in favor of animal sources in 1864 and then to the deliberate production and culturing of cowpox from infected animals. From then on, the pus used in vaccine was removed from cows' udders rather than from the arm of a child vaccinated a few days before.

Another factor contributing to declining vaccination rates was concern about the efficacy of the vaccine. Some children who were not vaccinated properly or who were vaccinated with spoiled material came down with the illness some years later. In 1820, the first cases of smallpox in young adults who had been vaccinated were seen, and several outbreaks in the 1820s destroyed the hope that protection would be permanent. Yet France's Central Committee on Vaccines did not accept the principle of mandatory revaccination until 1849, and revaccination was introduced very gradually.

Proper vaccination campaigns, involving the evaluation of results, were infrequently undertaken by local town physicians. Such efforts were especially rare in the surrounding countryside, as there was little pressure from regional authorities in favor of them. Physicians would spend perhaps just a few days per year in the nearby rural areas, regardless of whether parents could bring their children at those times. As a result, midwives became important as vaccinators. They had close relationships with both mothers and children and were available in most hamlets and villages. Once vaccinations were placed in their hands, results improved.[1] Rates of coverage were sufficient to prevent a major epidemic from developing until the 1860s. Smallpox struck Europe then not only because of wars but also because of the increased proportion of young people who had not received a new vaccination at age twenty. During the winter of 1870–71, 200,000 people died of smallpox in France. Sweden avoided this epidemic; there, vaccination had been required since 1816, and revaccination had been strongly recommended since 1853.[2]

In France, preventive medicine was floundering because of a lack of political will. Smallpox control marked the first great victory over an epidemic disease, but this victory was won only gradually during the last two decades of the 1800s, in a climate of mutual distrust and misunderstanding among the people, the doctors, and the authorities. Eventually, the victory over smallpox was achieved, thanks to the vaccination efforts of pioneering health departments of certain cities. These forward-looking organizations acted effectively before the requirements of the law of 1902. In 1902, vaccination became a legal requirement throughout France, and mandatory vaccination became the paradigm for preventive medicine.

Bacteriology and New Vaccines

Louis Pasteur proposed the new word *vaccination* in 1881 in honor of Edward Jenner and his famous procedure, which involved exposing a patient to an attenuated form of an illness in order to prevent the development of a serious case of the disease. Pasteur decided to model his work after Jenner's, and soon the new word came to apply to protective interventions for all infectious diseases, not just smallpox. In choosing this model, Pasteur changed the course of history and founded his own tradition. As Anne-Marie Moulin has written, "He reinvented the field of medicine in his era."[3]

Pasteur was struck by the fact that, in the history of smallpox vaccination, practice had preceded theoretical justification. He decided to proceed using this empirical approach and work on the basis of observations rather than theory. "People don't get measles twice, nor scarlet fever, plague, smallpox, or syphilis, so immunity must persist for some reasonably long time."[4] Before Pasteur, the vaccine was the illness itself, acquired from inoculation with the virus. For Pasteur, a true vaccine was something created by exposing infected material or the virus to the oxygen in the open air, attenuating it prior to inoculation. He began to work on producing such vaccines when he was admitted to the Academy of Sciences.

French scientists, led by Pasteur and his successors, concentrated their work on questions about immunization, while the Germans emphasized the discovery of new bacteria using new laboratory techniques such as staining the germs. For a long time, most French people believed that Pasteur was the inventor of the smallpox vaccine, especially since getting the vaccine was the only clearly understood obligation of the 1902 French public health law. To the French, this vaccine remained a powerful symbol of modern medicine for a very long time—so powerful a symbol, in fact, that use of the vaccine was continued in France until 1984, four years after the World Health Organization announced the worldwide eradication of the disease.

Pasteur's Laboratory Investigations

Pasteur discovered and isolated the germ that causes cholera in chickens and found that the germ is more similar to the plague bacillus than to the *Vibrio* that causes human cholera. He grew colonies of the new microorganism in chicken broth. He observed that cultures that had weakened over time could prevent illness among chickens subsequently injected with fresh, virulent strains of the germ. The analogy to vaccination was obvious, even though his protective material was artificially grown and was attenuated by exposure of the cultures to the oxygen in the air. These findings encouraged Pasteur to look for additional microbes whose virulence could be deliberately attenuated; he could then determine experimentally whether exposure to the weakened organism would provide immunological protection against other strains.

Next he chose to work with anthrax, an economically important disease of sheep. The disease tended to strike in massive, devastating epizootics (epidemics affecting herds of animals). After some trial and error, Pasteur succeeded in attenuating the germs by heating the cultures to 108 or 110 degrees Fahrenheit. He publicly demonstrated the efficacy of his immunizing procedure in May and June of 1881.

The next challenge was to prepare some type of new vaccine that would work in people; rabies was a priority. Pasteur was unable to identify or isolate the rabies virus, but he was able to grow it in the spinal cords of different animals. He noted that the virulence of the virus in this material diminished progressively when exposed to air: after two weeks the virus was no longer deadly for a dog in good health. Pasteur also observed that, when a series of injections of ever-increasing virulence was used, it conferred complete immunity against the bite of a fully rabid animal. In people, the incubation period between the bite and the onset of the symptoms of rabies is at least a month, and Pasteur saw the potential for effective vaccination during that interval. The successful attempt to save the life of young Joseph Meister confirmed this hope.

Between Jenner's and Pasteur's investigations, other doctors had tried to immunize people by injecting them with material taken from

Group vaccination against smallpox, from a cow in the garden, avoiding contamination from human arm to arm.

Painting by J. Scalbert, 1890. Courtesy of the National Library of Medicine.

syphilitic chancres, leprous nodules, gonorrheal discharges, plague buboes, and the blood of malaria patients—without success. Pasteur's work was different because he chose two diseases that occur in animals as well as humans, permitting more varied experimentation, and the germs involved had been identified and grown in vitro. A review of Pasteur's lab notebooks makes it clear that the source of his discoveries was continuous reflection coupled with persistence, rather than the lightning flash of genius, as the Pasteurian legend has it. Collaboration with the young physician Émile Roux was another critical factor in Pasteur's success. Roux eventually became the director of the Pasteur Institute (in 1904).

Vaccines against bacterial diseases—cholera, typhoid, and tuberculosis—were the first to be developed in the new series of vaccines. The cholera vaccine was developed by Haffkine at the Pasteur Institute in 1892. It was conceived and distributed as a vaccine for mass distribution, and thousands received it during the course of an epidemic in the Indies even though it was only partially efficacious.[5] Several years later, after the plague bacillus had been isolated by Yersin (in 1894), Haffkine returned to an active role in vaccine research, culminating in the plague vaccine in 1906. The two archetypal epidemic diseases, plague and cholera, had yielded to vaccine researchers in close succession. Almroth Wright, a British pioneer in the immunology of vaccination, tried to prevent pneumonia with vaccines he tested on South African miners. His greatest accomplishment was the introduction of typhoid vaccine in 1896–98. This vaccine had been tested on animals by Fernand Widal and André Chantemesse at the Pasteur Institute in 1888. World War I saw the first application of microbiology in military medicine, when the typhoid vaccine was given to British soldiers in combat.

Research on vaccination led to another important prophylactic procedure, serotherapy. In 1888, Roux and Yersin discovered toxins, showing that the filtrate from a culture of diphtheria bacilli could kill guinea pigs even though there were no bacteria present. In 1890, working in Koch's laboratory in Berlin, Behring and Kitasato demonstrated that animals could be immunized by the injection of nonlethal amounts of such filtrates. They showed that the blood of an immunized animal, transfused into another animal, conferred immunity. Thanks to Roux, the production of such immunogenic sera using horses became a practical reality in 1894, which paved the way for mass-produced treatments for diphtheria, dysentery, scarlet fever, and plague. In 1890, Behring and Kitasato developed a serum for the treatment of tetanus, which was widely used in World War I. With the exception of tuberculosis, syphilis, and malaria, the principal infectious diseases could now all be treated this way.

Tuberculosis: Feared, Resistant, and Romantic

During the second half of the nineteenth century, tuberculosis affected society more than any other disease. Unlike other epidemics that had occurred on a similar scale, tuberculosis was a "romantic" disease, giving rise to a new concept of interaction between the patient and a disease. The person with tuberculosis, weakened, was considered to have a heightened sensitivity.[6] The patient could have a "good death" because the slow progression of the disease made it possible to prepare oneself for the end, both personally and socially; one could say goodbye in the desired fashion.

Tuberculosis seems to have had an especially strong effect on society in the last two decades of the 1800s. Survey data support this view. In the *Annuaire Statistique de Paris* of 1889, Jacques Bertillon published a study of the frequency of the major diseases using data from the period 1865–87. He divided Paris into eighty neighborhoods according to their economic level. The average tuberculosis mortality was 460 per 10,000 inhabitants—558 per 10,000 among males and 364 per 10,000 among females. But there were marked differences among neighborhoods. The overall rate was as low as 182 in the eighth arrondissement and reached 628 in the twentieth arrondissement. During the twenty years under study, it seems that the differences among arrondissements became more pronounced. The shift in tuberculosis mortality, with outlying areas having the highest rates, was probably caused by urban renewal programs in the central city. Bertillon concluded that the disease was rarer in Paris than in the towns of Austria but more common than in London.

By the end of the nineteenth century, in the context of the distinctively French concern about the demographic future of the nation, tuberculosis was seen as an additional menace, becoming an obsession of the hygienists and the politicians. Physician members of the hygienist movement tried to publicize the problem. For example, a Dr. Armaingaud founded the Gironde Regional League against Tuberculosis, which led to the founding of the National League ten years later. These organizations played a role in education and prevention by distributing brochures and organizing conferences. An interim commission on the disease was established in 1899 by Waldeck-Rousseau; a permanent one by Combes in 1903.

Powerful antituberculosis activists used international comparisons to further the cause, demonstrating, for example, that the policies of Bismarck's government provided earlier and better intervention against tuberculosis—the first sanitarium opened in Germany in

1854—than the sporadic efforts in France. The tuberculosis death statistics cited by French hygienists at the end of the century were really quite inflated to serve propaganda purposes and for shock value. They gave a figure of 150,000 deaths, based on estimates obtained by recategorizing various causes of death. This is comparable to the number of deaths from cholera during the worst epidemic in 1854 and cannot have been an accurate estimate. Annual tuberculosis deaths certainly did not exceed eighty thousand.[7]

Research on the bacteriological aspects of tuberculosis made slow progress, in contrast to the more rapid advances seen in research on other infectious diseases. Still, knowledge of the anatomical and pathological aspects of tuberculosis had been growing since the start of the nineteenth century; Laennec's stethoscope was a crucial contribution to diagnosis by mediate auscultation.

In 1865, J. A. Villemin attempted to demonstrate that tuberculosis is contagious. Koch discovered the bacillus that causes the disease in 1882, giving hope that a vaccine might offer protection, along the lines of Pasteur's original successes. Koch worked toward this goal and thought at first that a useful treatment might be found in tuberculin, a material made from a concentrated filtrate of tuberculosis bacilli grown in broth. His experiments with this material led to a series of resounding therapeutic failures. Tuberculin can be used, however, in screening for the disease, since the skin's reaction to tuberculin is a measure of exposure to the bacillus. Put into practice by the Viennese doctor von Pirquet in 1907, the tuberculin test is still used today. X-rays were discovered by Karl Röntgen in 1895 and offered a new, low-cost way of confirming the diagnosis. In the end, however, no new treatment resulted from all this newly acquired knowledge, despite the hopes raised by Koch's tuberculin extract.

The discovery of the tuberculosis bacillus had a different effect on society. People with tuberculosis came to be seen as agents of contamination, who could spread the bacterium quickly to a very large number of people, especially in the big cities. They were a danger to society and to the elites. A false impression arose of the contrast between the "unhealthy, tuberculosis-ridden" city and the "pure air" of the "healthy" countryside. And tuberculosis was seen as the result of bad habits among the laborers, even in the factory, creating stigmatization of working-class behavior patterns. The link between alcoholism and tuberculosis was soon understood, leading to a distinction between those who were blamed for their own illness and those who were considered innocent victims of what was once believed to be a hereditary disease.

Compared to the situation in neighboring countries, the French campaign against tuberculosis made greater use of rhetoric concerning degeneracy, positing a link between the disease and venereal diseases, prostitution, and cabarets. In Germany, there was a greater emphasis on what Paul Weindling has called "the cleaning up of private life." In that country, tuberculosis was blamed on overcrowded slums, shared beds, dirty clothing, and alcoholism.[8] In England, it was caused by "poor food," "bad air," and "bad drink." In other words, poor nutrition, unhealthy living and working conditions, and alcoholism were thought to be responsible for the spread of the epidemic. The English reformers, typically, emphasized individual responsibility. They believed that cleanliness, moderation, and a calm life could prevent the disease. In England, there was little interest in heredity as a means of spreading the disease.

There was scant interest on the part of the French elites in hearing the demands of the workers. During a 1906 debate at the Permanent Commission on Tuberculosis, Brouardel explicitly rejected the idea that overwork might not only lead the workers to become dissolute but also directly contribute to illness by lowering resistance to tuberculosis infection. At the time there was an ongoing discussion about a proposed law that would establish the right to a weekly day of rest for employees. Brouardel would not allow improved resistance to disease to become an argument for the law. No matter what the philanthropists and leading hygienists wanted, political divisions were determining the national agenda for tuberculosis control. Members of the commission never contradicted the study of tuberculosis in the Calmette Dispensary in Lille, which showed that overwork contributed to the acquisition of the disease in 96 to 98 percent of cases. They were simply careful never to mention it.[9]

In France, the required reporting of cases faced two major obstacles. First was the totally uninterested or even skeptical attitude of medical practitioners toward statistical data. This attitude was the opposite of what had prevailed in the first third of the nineteenth century. In addition, there was concern about the confidentiality of medical records, which the profession was obligated to protect. The case report contained three main elements: the nature of the problem, the fatal outcome, and the circumstances of the injury or illness. To be conciliatory, the Committee on Public Health decided in 1892 to limit mandatory reporting to illnesses "posing a grave and present danger, in those cases where substantial benefit could be gained from having the information." This excluded tuberculosis because of the supposed hereditary predisposition and the lack of a lifesaving treatment for this illness. The level of concern about mandatory reporting of tuberculosis became manifest after passage of the 1902 law requiring the reporting of cases. Physicians were worried about losing their patients if the law were rigorously enforced and cases of tuberculosis were scrupulously reported by the medical profession. In response to pressure from physicians, the decree of February 10, 1903, made disinfection obligatory after any reported contagious disease, which made reporting seem more justifiable to many on all sides.

The continuing absence of effective therapy encouraged charlatans and increased the anxiety of doctors and families, who had to follow the accepted precautions that changed their daily routines and reached into their very homes. Theoretically the top priority, tuberculosis control efforts were not widespread. Public contributions quickly proved insufficient, and only government subsidies kept sanitaria and dispensaries open. (Organized charities and doctors' free medical services were available only for the indigent.) Tuberculosis screening was thus very limited, even where dispensaries existed. Moreover, dispensaries did not have the means to send infected patients for care in a sanitarium. The sanitaria, themselves in need of money, preferred patients who had been sent by community or regional health officials, who would pay.

The drive to open new sanitaria was not generally successful because politicians doubted the efficacy of a treatment that was not fully medical. Also, plans for sanitaria stimulated energetic protests from the communities in the projected locations. The local people refused to have a large group of tubercular patients in their midst.

Thus, until World War I, tuberculosis control efforts did not fulfill their potential, and very little was spent on such initiatives. The foundation of an effective network of dispensaries and sanitaria was to wait until the era of a more powerful central government and the Bourgeois and Honnorat laws of 1916 and 1919.[10]

The golden age of the hygienists in France was the time between the two world wars. New practices were imported by the Rockefeller Foundation and new funds as well—23 million "old francs." Visiting nurse services were established, and sanitaria were available for everyone of means. The medical profession took over the care of tuberculosis patients, on grounds that the disease was "morally debilitating." The patient was depicted as enfeebled and incapable of fend-

A Los Angeles sanitarium in the early twentieth century.
Courtesy of the National Library of Medicine.

ing for himself and above all as prone to vice (sexual excesses in particular). In the 1930s the image changed, and the patient was often presented as a victim of capitalistic exploitation, beset by poor working conditions and housing; he could resume his place in society once his health stabilized (as was happening more and more often). The disease wasn't completely irreversible, even if doctors hadn't found a real cure for it.

Researchers from the Pasteur Institute, Albert Calmette and Camille Guérin, developed the sole existing vaccine against tuberculosis. Between 1908 and 1921, they cultured and recultured (230 times) a certain strain of bovine tuberculosis on a potato soaked in beef bile. The successive generations, or "passages," of the germ differed in none of their original characteristics, except for an enormous loss of pathogenicity. Tests of the vaccine on laboratory animals and livestock were successful, leading to the first trial in people in 1921. In La Charité hospital in Paris, newborns were vaccinated by being fed three doses of a solution between birth and their seventh day. The statistics presented by Calmette at the League of Nations meeting in 1928 were eloquent. Fifty thousand infants received the vaccine between 1921 and 1926. Among vaccinated children living with a family member with tuberculosis, the death rate was 1.8 percent; among the unvaccinated it was 25 to 32.6 percent. The use of the BCG (bacillus Calmette-Guérin) vaccine, with its excellent safety record, was therefore recommended at the meeting.

The following year, however, the growing popularity of the vaccine was stopped short by the Lübeck disaster. BCG vaccine prepared in that town was given to 252 children; 73 died of tuberculosis and 136 developed the disease in chronic form. The vaccine had been accidentally contaminated by a strain of viable human tuberculosis bacilli. Despite reassurances that BCG itself was harmless, confidence among parents was shaken for decades. Many doctors complied with parental requests for waivers of the BCG vaccination requirement for schoolchildren. Even in France, the birthplace of the vaccine, the health authorities struggled against BCG vaccination until the late 1950s.

Fear of tuberculosis and the difficulties of treating it did not auto-

matically lead people to follow the recommendations and get vaccinated; many simply ran the risks of not being vaccinated. At the same
time, interactions between patients and doctors in sanitaria left behind a well-attested legacy of dissatisfaction with sanitarium treatment. The institutions, which were run according to very strict rules,
were places of suffering and worry dominated by the medical profession. Doctors made decisions on their own, without input from
others, concerning the advisability of massive, disfiguring operations,
such as surgically induced pneumothorax, thoracoplasty, and splen-
ectomy, with their painful results. After World War II, some sanitarium doctors fought to preserve their right to choose these "therapies"
in their institutions, even though antibiotics had appeared on the
scene, bearing the promise of a real cure without institutionalization.
After 1955, the world of the sanitaria suddenly collapsed, just as
France was constructing new ones at an unprecedented pace.

The cost of taking care of patients with tuberculosis became an issue after the end of World War I, at the time of the establishment of
the "welfare state" in Western Europe. It was such a problem that,
starting in 1929, the civil service in France added medical exams as a
requirement when recruiting for competitive positions. Tuberculosis
patients and former patients thus tended to end up working in sectors of the economy that lacked job security, an important development in view of the economic crisis of the 1930s. Tuberculosis also
placed a tremendous burden on families; adding the cost of work-
days lost because of the illness, the economic effects of tuberculosis
would truly be phenomenal.

The Twentieth Century: New Vaccines despite Theoretical
Uncertainties

Theoretical uncertainties were becoming more evident. In the late
nineteenth and early twentieth centuries, progress in microbiology
had provided tools such as prophylaxis and vaccines, but there was a
growing awareness that these advances had been obtained without a
detailed knowledge of the mechanisms of immunology and that
more basic knowledge would be useful. For example, live vaccines
were expected to be more effective than those based on killed mi-

croorganisms because the living organisms should provoke a greater immune response. On the other hand, there was suspicion that a live vaccine would have a greater chance of causing side effects and could even regain its virulence. Empirically, it seemed that a germ became attenuated as it continued to circulate within a species, yet it regained its vigor when it passed from one species to another. The answer to an important theoretical problem thus remained unknown; the attenuation of germs in vaccines that had been developed in animals might be reversed when the same vaccine was used in people.

The issue of whether to use killed or attenuated vaccines came up again in the context of research on poliomyelitis. Jonas Salk made an injectable polio vaccine from inactivated virus; clinical trials were conducted starting in 1952, and the vaccine was readily available by the start of the 1960s. On the other hand, Albert Sabin seemed to obtain excellent results with living, attenuated oral vaccine, which was tested in 1957–58.

In the case of influenza, there were other difficulties. The Spanish flu killed millions at the end of World War I, motivating an expansion in research on the subject, but investigators were stymied by their inability to see the virus. In France, the flu raged from June 1918 through the first few months of 1920. Records available from the seventy-seven départements that had not undergone wartime occupation showed that the disease killed 137,000 civilians and 30,000 in the military. This epidemic exceeded the mortality of the 1854 cholera outbreak, which had killed 150,000. The scourge killed three hundred people a day in Paris in the autumn of 1918; it ultimately claimed at least 25 million lives worldwide.

Eventually, in 1940, two different flu viruses were isolated. Until the end of the 1950s, some people considered viruses to be like miniature bacteria, while others thought that viruses were a non-living chemical substance; accordingly, there were two different approaches to research. In 1941, however, Macfarlane Burnet discovered that influenza virus could be grown on the membranes of fertilized chicken eggs. This discovery made research much easier and led to the preparation of the first flu vaccine, based on two different viral strains, which was tested on a large scale in 1943–44.

Influenza is less predictable than other diseases: its behavior changes from one epidemic to the next because of mutations. Thanks to improved surveillance in China, where many new strains arise, and the isolation of "harbinger strains," which evolve at the end of one epidemic and form the basis of the next, the vaccine can be better adapted to the virus that causes the next epidemic. Despite the technical ability to make vaccines, most people consider the flu a minor ailment, so the vaccine has been underutilized. This "minor" ailment killed at least sixteen thousand in France in 1968, despite the availability of antibiotics for the bacterial infections that often accompany flu. It wasn't until the 1980s that the use of the flu vaccine became widespread in Western countries, finally gaining acceptance even though it must be injected annually.

The list of twentieth-century medical achievements must include the defeat of rubella, or "German measles." Known since the late 1700s, rubella was a commonplace cause of fever and skin lesions. It was considered unimportant until 1941, when an Australian study discovered its effects on the fetus. Rubella damages the heart, brain, and inner ear of the unborn baby and causes cataracts. The German measles virus was isolated in 1961 by American researchers, two years before a huge epidemic swept Europe. The epidemic spread to the United States in 1964 and affected twenty thousand newborns. The first clinical trials of the vaccine took place in 1967–69, and the vaccine reached the market in Europe in 1970. It was so successful that, a few decades later, the disease is practically unknown in Western countries. A healthy baby is such a priority for parents that even the slightest risk has become unacceptable, and vaccines have become widely accepted preventive measures.

Similarly, parents have warmly accepted the vaccine produced from the measles virus strain that Enders isolated in 1960. For three decades now, the practice has thus been to vaccinate jointly against measles, rubella, and mumps.

Objections to Vaccination

Inoculation has not always been so widely accepted. In many parts of the world, the appreciation of smallpox inoculation was tempered by concern about accidental mishandling of the material that can lead to raging epidemics. People have long recognized that they run a risk in inoculating their children against smallpox. Once Jenner's vaccine became available, the risk became very small, at least from the 1820s onward. On the other hand, the animal origin of the vaccination fluid has made many parents hesitant to accept the procedure. Some have even preferred to take the risk of inoculating their children with smallpox itself to be certain that their child will forever be immune to the disease.

It is clear that the transmission of syphilis—cases of the disease in fifteen- to twenty-year-olds—have not encouraged use of the vaccine against that disease. In other cases, poorly preserved vaccines of various types have conferred no protection, leading many to resist vaccination where this has been observed. Moreover, the medical community sometimes has insisted that vaccinal fluid (for smallpox) simply cannot be implicated in the transmission of another disease, syphilis in particular, despite the observations of the population to the contrary. Similarly, professors of medicine in France were slow to accept the concept of revaccination, waiting until 1849 to do so officially, although it had been common knowledge since the late 1820s that vaccination would not necessarily confer lifelong protection.

The unsuccessful attempts to vaccinate against syphilis and tuberculosis in the nineteenth century did little to encourage enthusiasm for vaccination in general. But the decisive factor in the acceptance of vaccines was usually the way in which vaccination campaigns were administered. In authoritarian regimes, they were seen as a means of control and perhaps as a way of identifying the young men in the country, something not very reassuring to parents.

Organized Political Opposition

Opposition movements, organized to resist vaccination, arise mainly when political parties are too weak to achieve the goals of the people. Stricter government vaccination requirements tend to stimulate a more active movement. In Britain, the dashed hopes of the Gladstone administration's reform campaign led to an alliance involving a vast array of nonconformist movements, including opponents of vaccination, vivisection, temperance policies, militarism, and slavery. The same group also fought against the suppression of alternative medicine, as represented by homeopathy.[11]

The anti-vaccination movement was strongest in the Protestant countries of northwestern Europe, that is, in the countries that had tried to establish a truly mandatory vaccination policy. The policy was opposed for several reasons. Beyond a general suspicion of those who made their living with the lancet, a deeper issue was often raised: Was humanity impinging on divine prerogatives? Was the lancet, perhaps, the enemy of the cross?

Most members of the anti-vaccination movement presented themselves as representatives of good common sense. They were supposedly fighting to preserve the rights of the individual to self-determination, against the doctors' and vaccinators' lobbies (which had the advantage of state support). The opposition wanted to see a more democratic approach, with safeguards for individual free choice, as opposed to an approach based on the coercive power of the state. Britain and Germany, especially, saw a large volume of publications decrying vaccination and an equally large stream of pro-vaccination writing.

The opposition thought it best to wait for proof of the safety and effectiveness of vaccination. Doctors had taken decades to accept the need for revaccination and had long denied that syphilis could be transmitted by arm-to-arm vaccination; these blunders gave weight to the arguments of those opposed to vaccines. And the vaccination procedure itself was controversial; the lancet was a violation of the person and also introduced materials directly into the bloodstream without intervening digestive processes to break them down. Some people blamed these foreign materials for a host of ill effects, like tooth loss, hysteria, masturbation, hemorrhoids, scrofula, tuberculosis, cholera, and syphilis. Vaccination was "contrary to the laws of nature"—and contrary to logic as well, since a poison was being introduced as a protective measure. A person in good health was being contaminated to afford protection against a potential illness at some time in the future. Furthermore, the introduction of foreign material into the body seemed to go against the new ideas concerning asepsis.

Vaccines now came directly from infected cows, provoking opposition due to the fear of mixing the blood of two different species. Political cartoons showed "bovinized" people: half man, half beast. Scandalized by the apparent transgression of the interspecies boundary, the religious deplored bovine-derived vaccines, which they saw as an example of the degeneracy of the human race.[12]

As this controversy played out in the late 1800s, the various parties involved often took on the roles they had had in the days of the cholera epidemics. Those who had supported quarantines, such as Koch in Germany or Proust in France, were on the side of vaccination. Those opposed to vaccination included many hygienists, who emphasized the importance of the environment. For them, smallpox was but one of the illnesses caused by the lack of cleanliness, and these illnesses could all be dealt with in a single program of sanitary reform along the lines of Chadwick's midcentury reforms.[13] Some people felt that everyone ought to live in harmony with nature in a clean environment and that harmonious individual everyday behavior was the best preventive measure against illness. Moreover, the political will to adapt quarantine systems to the new challenge of smallpox could also lead to more draconian isolation measures, as it did in Leicester.[14]

The anti-vaccination movements, whether in Britain, Germany, or Sweden, drew membership largely from the working classes and sometimes the tradesmen. As to religion, members of these movements tended toward dissent from the established Protestant churches, such as Quakers in England, Baptists in Sweden, and Pietists in Württemberg. Faced with successful tactics on the part of various anti-vaccination movements, the British Parliament passed laws in 1898 and 1907 lightening the restrictions imposed on parents who sought to avoid vaccinating their children.[15] In France, the Ligue Universelle des Antivaccinateurs, with its coalition of scientists, intellectuals, doctors, and celebrities, was probably responsible for the failure to pass the Law Liouville in 1880, which would have made vaccination compulsory. As cases of smallpox in European countries grew rarer in the first half of the twentieth century and then disappeared, vaccination requirements were dropped.

But has the disease really been eradicated for good? In response to concerns about terrorism, the United States has geared up to produce millions of doses of smallpox vaccine once again.

In the second half of the twentieth century, more vaccines became available and they became more widely used. Today we have vaccines against diphtheria, tetanus, poliomyelitis, tuberculosis, measles, German measles, whooping cough, mumps, and hepatitis B, and research is continuing on vaccines against malaria and sexually transmitted diseases such as AIDS. Opposition to vaccines, however, has by no means vanished, even though their practical efficacy has been amply demonstrated. Sometimes opposition arises in smaller countries as a manifestation of resistance to the hegemony of the world powers. More generally, it arises from an unwillingness to run an individual risk for the benefit of the group.

Additionally, people are well aware that vaccines are not perfect. The polio vaccine is known to cause one case of paralysis per 500,000 vaccinations, and the rabies vaccine has always caused damage to some recipients.[16] There are many unanswered questions about the relationship of vaccines to subsequent disease. Does vaccination cause a temporary diminution of immunity, leading to easier infection? Does insufficiently attenuated virus undergo an increase in virulence? Could it be that vaccination activates virus already present, lying dormant in the body? Or could the problem really be the use of animal tissue in the manufacture of vaccine? Perhaps vaccination does stimulate the immune system, while sometimes causing flaws in it.[17]

Beyond their overall effectiveness in preventing disease, vaccines were a necessary precondition for the growth of trade on the global scale. After all, the systems of partial or limited quarantines ultimately presented obstacles to international traffic, just as the old quarantines and blockades did. The protection afforded by vaccines made travel possible, at least for those who had the means: the people of the rich countries.

7

The list of diseases preventable by vaccination grew markedly in the developed countries, in the late twentieth century in particular; this growth was paralleled by the advent of therapeutic advances such as sulfa drugs and antibiotics. A great number of infectious diseases could now be treated effectively, including tuberculosis and syphilis. At the same time, those countries developed social welfare programs that made medical care generally available throughout the population. Enormous economic growth was observed after World War II, raising the standard of living at all levels of Western society. All of these developments help to explain the extraordinary improvements in life expectancy that were enjoyed throughout the period. For a golden moment in the mid-1960s, it was possible to imagine that infectious diseases might some day be eradicated completely, although at the time only smallpox had succumbed to dedicated efforts.

Bacteriology's Successes: Sulfamides and Antibiotics

Paul Ehrlich was a physician and the nephew of a dye manufacturer. Ehrlich demonstrated that specific tissues could be colored by specific dyes. For example, cells of the nervous system were stained by methylene blue. In 1882, Ehrlich showed that Koch's tuberculosis bacillus could easily be observed if stained with fuchsin red, opening the way for a diagnostic test that was to become standard. A true founder of pharmacotherapy, Ehrlich went on to use coloring agents as pharmaceutical treatments. He had realized that the colors not only became affixed to microorganisms, they also killed them. A red dye was useful for treating trypanosomiasis, the disease better known as African sleeping sickness; the dye was thus called trypan red. In 1909, after having tested more than six hundred different dyes, Ehrlich's team noticed that the 606th possessed exceptional curative properties in rabbits infected with syphilis. It was thus discovered that arsenical compounds were useful in treating the venereal disease, and dye 606 became commercially available under the name of Salvarsan, followed by the less-toxic Neosalvarsan in 1912. This product remained the mainstay of syphilis treatment until penicillin was produced on a large scale in 1945.

Ehrlich's work on the staining of microorganisms was a breakthrough in pharmaceutical research. In its wake came Gerhard Domagk's research (1932–35) on a brick-red dye called Prontosil. He discovered that it saved mice after they had been infected with *Streptococcus* bacteria.

At the Pasteur Institute, Ernest Fourneau had been assigned by Roux to direct a laboratory dedicated to the development of new pharmaceutical chemicals. One of his teams looked for the active ingredient of dyes, the actual component that was killing the bacteria. They proved that, after dyes were metabolized, a simple, colorless molecule was the fatal agent. It was called p-aminobenzene sulfonamide. After this discovery, it became possible to find related molecules and to maximize both the safety of the medicine and its efficacy against germ targets. The sulfamide group of compounds proved to be new and useful weapons against coliform bacilli (some of which cause dysentery) and also against *Staphylococcus, Pneumococcus, Meningococcus, Gonococcus,* and some streptococci—specifically those causing the terrifying puerperal fever that was killing so many new mothers in those days. The use of sulfamides lowered death rates immediately: for example, the death rate in pneumonia cases fell from 20 percent to 5 percent after the new medicine was introduced. Sulfamides were of no use, however, in treating tuberculosis, typhoid, syphilis, or various nonpuerperal types of staphyloccal infection.

In the late 1800s, several microbiologists (such as Pasteur and Joubert) had observed the competition between different microorganisms. The other productive line of research in antimicrobials arose from this observation. In 1928, while growing a strain of *Staphylococc-*

cus, Alexander Fleming noticed that no staph grew near a green mold, *Penicillium notatum,* an accidental contaminant of the culture. Fleming realized that the mold was secreting a substance that prevented bacterial proliferation. He called this substance penicillin and published his discovery in the *British Journal of Experimental Pathology* in 1929. Ernst Boris Chain and Howard Walter Florey began research on penicillin in 1939. They isolated and purified it and then measured its effectiveness and toxicity at various doses. Their work led to the industrial production of the first antibiotic in 1943. The importance of this discovery was immediately recognized; the three researchers were awarded the Nobel Prize in 1945.

The next few years saw massive resources directed toward antibiotic research in the United States. Discovery of additional antibiotics, derived from fungi, soon followed. Selman Waksman discovered streptomycin in New Jersey in 1944, and aureomycin was discovered in 1948, followed by the cyclines. Other antibiotics were obtained at least in part by chemical synthesis, like chloramphenicol, which was made for the first time in 1947 and turned out to be very active against germs causing typhoid and dysentery.

Soon after it was discovered, streptomycin was tested in a sanitarium in Minnesota; the encouraging results were disseminated to the scientific community the following year. The experiments were repeated, and the U.S. National Tuberculosis Association announced the findings at its annual meeting in 1946. For the first time, a truly effective treatment against tuberculosis—a goal since the late nineteenth century—had become available. Very soon streptomycin became widely known and used in regions of Europe under postwar reconstruction with a substantial American presence. For example, it was tested at the Percy Hospital in France in 1947.[1] Although streptomycin was very effective against tubercular meningitis, the antibiotic could provoke bacterial adaptations that led to relapse.

This initial finding of antibiotic resistance could be handled. Paramino-salicylic acid had been shown to be effective against tuberculosis in 1943–44 experiments by the Swedish researcher Jorgen Lehmann, while isoniazide was proven effective in 1952 by Bernstein and Robitezk in the United States. The classic treatment of tuberculosis for several years was a triple therapy using streptomycin, paramino-salicylic acid, and isoniazide. The effectiveness of this triple therapy was remarkable. A resurgence of tuberculosis had been feared in the wake of wartime deprivations, yet tuberculosis continued to decline in the five-year period from 1949 to 1953 at exactly the same rate as had prevailed over the previous twenty-five years. Despite the opposition of sanitarium doctors, antibiotic usage expanded quickly. This expansion was stimulated by newer antibiotics, which dealt with the problem of resistant organisms arising from insufficient or interrupted treatment with earlier antibiotics. The newer generation included neomycin (1949), ethionamide (1956), and ethambutol (1961), followed by rifampicine, discovered in 1966. Rifampicine was particularly useful, since it reduced the duration of the contagious period.

Surgical interventions, such as the removal of pulmonary lobes, were not abandoned immediately. On the contrary, antibiotics permitted better surgical outcomes, and it took a long time for sanitarium doctors to appreciate the unforeseen efficacy of antibiotics. At a sanitarium at Saint-Hilaire-du-Touvet, for example, the annual number of lobe removals grew sharply until 1955, and then the operation vanished in the early 1960s.

Victory over Tuberculosis

Those who considered tuberculosis a social disease were not persuaded to abandon the idea by the effective new therapy. After all, since the 1960s the poorer social classes have been made up of immigrants, among whom tuberculosis rates are substantially elevated. In France in 1970, immigrants from sub-Saharan Africa made up 3.5 percent of tuberculosis cases but only 0.1 percent of the population. Immigrants from former French colonies in North Africa were 11 percent of cases and 1.9 percent of the population. Ten years later, the tuberculosis incidence rate was 23.7 per 100,000 in France, while it was 344 for African immigrants and 138.9 for those of North African descent. Migrant workers are thus perceived as importers of a disease that has been beaten in rich countries but ravishes the poor countries, with annual infection rates 20 to 50 times higher in the latter. Pierre Guillaume summed it up well: "Thus, this illness was formerly an indicator of poverty in developed nations, while now it is a symptom of underdevelopment."[2]

Of course, tuberculosis bacilli that resist antibiotic treatment have emerged, but it is important to recognize the limited magnitude of the resulting difficulties. Tuberculosis death rates fell from 100 per million in 1965–70 to 11 per million in the late 1990s. The prevalence rate dropped from 100 per 100,000 in 1955 to 15 per 100,000 in 1995, although the disparity between French people (9.8) and immigrant groups in France (66) remains substantial. This large gap is indicative of the precarious health status of immigrants and of their unfortunate living conditions. If unconfirmed cases are included, a small increase in the number of cases has been observed since 1992, and this rise is attributed to immigrants and to people with AIDS.

The reception of immigrants is influenced in contradictory ways by their high tuberculosis rates. On the one hand, society is moved by their plight, and today more than ever it is unthinkable to ignore the needs of a person whose health is in danger. On the other hand, the foreigner has become a potential means of bringing back the epidemics of the past. Over the centuries "the stranger" has been seen as a danger to the public health, and this view has thus been revived. People with AIDS are affected by weakened immune systems, and their health status probably hasn't helped their social integration. In any event, despite the arrival of multi-drug-resistant tuberculosis, the number of cases of the disease is still shrinking at a rate of 6 percent per year in France, if only confirmed tuberculosis cases are considered.

Overall, the number of deaths from infections fell by 20 percent between the two world wars; it fell another 50 percent between 1945 and 1950 and 50 percent again from 1950 to 1960. (The very sharpest of the declines was from 1945 to 1947, with the initial introduction of antibiotics.) Vallin and Meslé estimated that life expectancy increased by nine years between 1945 and 1965 because of the near-disappearance of fatal infections alone.[3] During those twenty years, antibiotics proliferated, complementing each other, and their joint effects offered unforeseen possibilities of controlling most infections. In the history of humanity, or at least the history of the rich countries, the twentieth century was the era that finally saw an extreme reduction in mortality from infectious disease.

Industrialization and the Expansion of Demand

Most germs could be destroyed by the wide variety of antibiotics available. The new medicines were used extensively, first in hospitals and then in general practice, starting in the late 1950s. The worldwide consumption of antibiotics was measured in the millions of tons. Production on this scale involved industrial fermentation vats of remarkable proportions (50 to 200 cubic yards), as well as automated filtration and extraction equipment. The enormous sales growth of antibiotics could not have occurred without the spectacular decline in manufacturing costs due to economies of scale, which made retail prices affordable. The retail price of streptomycin in 1954 was 1 percent of the 1946 price. Later, partially due to affordability, consumption of the medicines increased to an unprecedented extent. For example, in France between 1970 and the late 1990s, the per person annual expenditures on antibiotics increased tenfold adjusted for inflation, while the average price for pharmaceuticals increased only one-third as much as a general price index of consumer goods.

The increasing availability of antibiotics of ever-higher quality and purity—at declining prices despite increased investments in research—could only have been achieved with a restructuring of the traditional pharmaceutical laboratory. There were two thousand laboratories in France at the end of World War II; only 700 remained in 1964, and 450 in 1970. The top fifty companies in the 1970s took in 80 percent of the revenues of the industry.

Synthélabo is a large French pharmaceutical firm, which was formed in the 1970s as the result of an alliance between Dausse, Robert, and Carrière.[4] In 1973, L'Oréal acquired a majority stake in Synthélabo, clearly showing that L'Oréal's president, François Dalle, understood the value of getting involved in an industry with such excellent prospects. But he also sought access to the scientific strengths of such a company, since he could use those resources to assure the safety of L'Oréal's cosmetic products. Such an enormous corporate merger was made possible by the remarkable commercial success of cosmetics, built upon the desire to lavish attention on the body combined with the financial ability to do so.

The acquisition led to the availability of enormous capital reserves, which were indispensable to corporate growth in the 1970s. Thanks to the infusion of cash, Synthélabo's research and development division was able to hire a constellation of new researchers of international stature. Giuseppe Bartholini, formerly the director of the world's top biochemical pharmacology group at Hoffmann-LaRoche, was appointed the director of the division in 1975. Little by little, Synthélabo absorbed other important research-based companies, like Delagrange and Delalande, and became a major international pharmaceutical corporation. Investment in research went from 6–7 percent of revenues in 1970 to 15–20 percent in the 1990s. The game was now being played for serious financial stakes.

As the pharmaceutical industry grew, the state got into the business of regulating medicines in the marketplace. The government is on one side of negotiations as agreements are reached about putting new drugs on the market, importing medicines, and including drugs in health insurance benefit plans, especially plans provided by the government. On the other side of the negotiating table is usually a multinational pharmaceutical industry giant, and they are very powerful firms.

Government Programs

In the 1930s and 1940s, most industrialized countries instituted medical insurance and social security plans, systematizing reimbursement for medical care and for prescription drugs. This step was a huge leap forward for the health of the population. It also ensured the prosperity of medical practitioners and large pharmaceutical companies.

Mutual insurance companies had existed since the mid-nineteenth century, offering insurance policies that would pay benefits to replace wages that workers had lost due to illness. Some large industrial firms had long had paternalistic attitudes toward their workers and offered free medical visits and drugs. Laws initially passed by

Bismarck's government in 1883 and 1884 came within the traditional scope of aid to the poor; their primary goal was to prevent indigence. Little by little, government plans in Germany expanded in scope, as amendments of the law moved toward a greater emphasis on medical insurance for workers, until by 1892 all insurance funds offered medical and drug coverage to the families of the insured.[5] In the twenty years between 1891 and 1911, the average annual medical expenses for each insured person almost doubled, from 7.6 to 14.1 marks. Inevitably, pressure arose to keep medical expenses in check. This pressure became manifest in restrictions on physicians working for government pay, who might, for example, be required to limit their prescriptions of the most expensive medicines. But there were democratic trends as well: under the new plans, workers not only obtained medical attention with ease but also even had access to specialists.

Between the two world wars, the German model was adapted to the institutions and cultures of other developed countries. In France, most large companies had established medical plans for their employees, and sometimes for their dependents, since the time of Napoleon III. But the first insurance plan covering everyone was approved in 1930, after extensive parliamentary discussion; it was called *Assurances sociales*. Some politicians feared an explosion in demand for medical services. Doctor's visits, most medicines, and even hospital stays were to be reimbursed at 80 percent of the cost to the patient. Louis Loucheur, minister of health in 1930, planned for four thousand new hospital beds in the Seine region, and another ten thousand for short-term use, to handle the huge load of patients that the new law would undoubtedly generate.

Restrictions were put in place to limit the demand. For example, benefits were extended only to those who had contributed to the plan for at least sixty days (out of a total of sixty-five possible working days) during the three months before an illness. This requirement was a real obstacle in the context of a difficult job market. As to hospitalization coverage, total daily benefits were limited to about 40 percent of the average salary of the Parisian worker, which could be as little as one-third the salary of a well-paid worker. Moreover, a five-day delay in payments temporarily left the patient without any compensation. As to medicines, the 80 percent reimbursement level was limited to 25 francs per prescription; beyond that amount, the percentage decreased. And plans set the reimbursable amounts sometimes unrealistically low—often lower than the fees ordinarily charged by doctors or hospitals. Thus, at the end of World War II, the real rate of reimbursement was scarcely 50 percent.

Other factors were at work. Mutual insurance fund directors expressed a concern about the citizenry's loss of free choice, as well as the devaluation of foresight and planning. These concerns were addressed in the law of 1930 by retaining the nineteenth-century decentralized structure of health insurance and by permitting any interested parties to set up insurance funds. Consequently, some funds were created under the auspices of Catholic organizations, some were set up by employers, some were set up by private insurance companies, and some were sponsored by trade unions; all were competing for policyholders. In addition, regional "neutral plans" were created for those with no preference.

By 1944 there were 589 main insurance providers offering medical and maternity benefits. There were "old-age funds" on top of that, as well as four hundred "children's benefit funds." To preserve order in this huge number of competing programs, the government retained some central functions, such as registration of plan members and collection of premiums.[6] The mutual insurance companies emerged as the dominant player, with 20 percent of all plan members. Such companies tended to recruit customers in the traditional middle classes, so labor union plans readily portrayed the "mutuals" as outsiders, alien to the concerns of most workers—a useful view to promulgate in view of the fight for dominance. World War II brought complaints that the law of 1930 was "too interventionist" (according to doctors and small and large businessmen) and "not interventionist enough" (according to labor unions and some top government officials).

The impetus for reform came from outside. The Beveridge report, published in 1942, changed the frame of reference by proposing social security programs involving universal coverage and a single payer. Protection ought to extend to everyone and should cover all risks that might affect the ability of individuals to provide for themselves and their families. The single-payer system implied a melding of public assistance with private insurance and a uniform benefits schedule, in which all care would be free to patients and all reimbursement levels identical. Britain established its National Health Service in accordance with these principles.

The International Labor Organization took an idea of Franklin D. Roosevelt's and declared in May 1944 that member nations would work toward true social security systems, as soon as the war was over. The Comité Français de Libération Nationale expressed demands for "complete social security coverage, guaranteeing all citizens a means of subsistence whenever they become incapable of obtaining it for themselves by work."[7]

The political situation at the time of the liberation of France was favorable to such goals because of the weakened power of employers and the loss of legitimacy in the insurance industry, which had compromised with the Nazi puppet government at Vichy. Also important was the power of the French Communist Party and the agreement between the socialists and others on an ambitious social program. Pierre Laroque, the "father of French social security," stated in January 1945 that "the British plan is all the rage." For Beveridge, the entire country was really an organization for the purpose of freeing individuals from need, through economic and social programs. "Welfare state" policies included full employment, high salaries, and redistributive taxation and spending. Finally, the laws of October 1945 established a single system that improved reimbursements by providing total coverage in the case of hospitalization and removing limits on daily charges, while limiting unpaid days to a total of three. (Not all plans were unified, however, as workers in certain industries, such as farming, railroads, and mines, continued in separate programs.)

Unlike the previous situation, benefits could now be brought in line with the income level and lost wages of the insured, at least up to a certain level. Moreover, starting in December 1945, there were no longer reimbursement limits of 80 percent and 25 francs per prescription. For chronic illnesses, long-term care insurance covered all costs for up to three years. This provision was especially important for those with tuberculosis. For their part, doctors accepted the state's regulatory role in setting fees, in exchange for the reimbursement being rather generous.

Government expenses for health care went up 25 percent per year, and by 1947 the medical care insurance funds (including long-term and maternity care funds) were running enormous deficits. Some of the costs had to do with the postwar baby boom, but there were two other factors as well. One was the poor general health of the population after the miserable years of Nazi occupation; individuals were at last turning their attention to their health problems. The other was the arrival of antibiotics, for modern medicines carried higher costs. The cost of medications had an effect on hospitals; data on some public institutions in Lyon show a doubling of expenditures on medicines between 1947 and 1948 (adjusted for inflation) and an additional rise of 22 percent between 1948 and 1949.[8] Even these factors combined do not entirely explain the rising expenditures; social security programs were favoring ever-increasing access to services, as had been the trend ever since the late 1930s.

Henri Péquinot had a different explanation. He taught a course at France's prestigious École Nationale d'Administration in 1952, in which he expounded a new thesis about the growth of health care costs, keeping in mind that for the first third of the nineteenth century, rules were in place to restrict inappropriately long hospital stays, excessive "sick days," and spa cures. According to Péquinot, "The level of health-related expenditures is basically a function of the 'medicalization' of the country, that is, the current level of medical technology and the psychological expectation that a population has with respect to the consumption of medical services."[9] He also observed that the cure of any illness can't make any contribution to the economy because the illness is immediately replaced by another ailment that had not been correctly treated previously.

Péquinot's theories are the polar opposite of the traditional view, which sees health expenditures as the result of poverty. Instead, the higher the standard of living and the better the sanitary conditions,

the higher the consumption of health care—and the tendency is always to grow. Péquinot's analysis was insightful, and his expecta-
tions have been essentially confirmed by subsequent experience.
Hospitals have had to adapt rapidly as the indigents of earlier times
have been replaced by an increasingly well-off population seeking
the benefits of medical progress. In addition, there is competition for
hospital patients, who are choosy about the location of services, the
comfort of hospital rooms, and the attitude of the staff. The modern
argument about how to control public expenditures for health care
began in the postwar period and has never abated. The real obstacle
to any cut in benefits is the pressure exerted by a population that has
overwhelmingly chosen to devote a large portion of its income to the
care and maintenance of the body. French society, to pick just one example, reacts negatively to any delay in the marketing of newly developed pharmaceuticals. This consumer pressure is strong, even
though the eradication of infectious diseases, which seemed within
reach in the 1960s, seems much less achievable decades later.

There have been great improvements in life expectancy at birth.
French society is typical: in 1950, life expectancy was 63.4 years for
males and 69.1 for females; in 1970, the figures were 67.6 and 75.2, respectively. The gains result from institutions providing better access
to better medical care and the provision of guaranteed retirement income for all. The gains are also due to antibiotics—a scientific breakthrough—and strong postwar economic growth, which made possible the financing of all the medical and social programs of the period.

8

In the early 1970s, it was still possible to anticipate the end of infectious diseases, as the ultimate result of work that emerged from nineteenth-century beginnings. This daydream was soon abandoned in the face of infectious disease rates that improved very little in the 1970s. The stagnant rates were the result of antibiotic-resistant germs, which posed an especially serious problem in hospitals. In addition, so-called emerging infections came to light, including AIDS, hemorrhagic fevers such as Ebola, and fatal pulmonary disease caused by hantavirus, carried by small rodents in the wild. The disappointment was in proportion to the hopes that had been raised by the earlier faith in progress.

Resistance and Emerging and Re-emerging Infections

The proportion of all deaths that are due to infection has remained fairly constant since the 1970s, particularly among the elderly. There are several reasons for this lack of improvement. First, antibiotics are sometimes used unnecessarily, while at other times the patient takes a dose inadequate to kill off the bacteria for which the drug was prescribed. The inappropriate use of antibiotics leads to resistant strains, rendering the usual treatments useless. This problem is naturally most common in hospital settings, where the patient population tends to be the sickest and many are dying. Yet the prevalence of antibiotic-resistant bacteria is highly variable from country to country.

In France, the awareness of this problem resulted in the establishment of an Infection Control Committee as early as 1973; hospital-based infection control committees were instituted in 1988, and a national policy on hospital-acquired infections was adopted in 1994. This slow development does not imply a lack of activity on the part of hospital executives and staff for twenty years; however, the development of a specific committee was slow in coming.

The approach of the French, and of Europeans more generally, was different from the approach taken in the United States. European anti-infection efforts were an echo of the old hygienist movement: the emphasis was on interrupting the chain of transmission of microorganisms from person to person and on destroying the germs in the patient and his living quarters. Infection control responsibilities were thus divided among various groups, each pursuing its own goals as far as resources would permit. The goals included the correction of individual behavioral errors, such as lack of vigilance, imperfect hand-washing, or failure to avoid contamination by disinfecting the surroundings properly.[1]

While European infection control activities focused on altering individuals' behaviors, the American approach was based on the model endorsed since 1970 by the Centers for Disease Control and Prevention in Atlanta. Here, the epidemiological orientation involved the consideration of a population in its entirety, as a system. It was important to examine risk factors and identify risk groups. This population-based approach necessarily relied on a health care system that would give priority to the efforts of infection control committees and to the establishment of specific hospital services staffed with specialized personnel. Clearly, the emphasis was on collective rather than individual responsibility, leading to the examination of dysfunctional aspects of systems and organizations. Outbreaks are studied using a case-based approach. In the mid-1980s, an evaluation of these measures showed that they had reduced the risk of hospital-acquired infections by one-third.

Whether the European or American approach to hospital control activities was adopted, a substantial change had taken place. The problem had previously been seen from a microbiological perspective, involving the discovery of new pathogenic organisms as targets for destruction. Now it was a matter for epidemiological investigation focused on the effects of interventions on populations. Previously, a distinction had been made between infections due to endogenous bacteria (those "native" to the patient) and those due to exogenous bacteria; only the latter were considered hospital acquired. But now epidemiologists were considering any infection during a hospital stay to be nosocomial, or connected with treatment. In light of this new definition, the search for a guilty party responsible for an infection was abandoned; responsibility was shared.

Data indicate that the average rate of nosocomial infection in France was 7 percent in the 1990s. Ongoing surveillance of nosocomial illnesses, however, now indicates a sharp increase in these ailments. Surveillance data help identify the sources of hospital-acquired infections because comparisons can be made not only between hospitals but also between different units within the same hospital; the practices of individual health care workers must be examined. Thanks to the new perspective, the efficacy of an intervention is now evaluated in terms of a reduced infection rate rather than a reduction in the number of microorganisms recovered from a sampling of the hospital environment. This is especially important, since pathogenic organisms may possess varying resistance to antibiotics.

A study conducted in 1992 showed that resistance to methicillin—the principal defense against *Staphylococcus aureus*—varied to an enormous extent. In the northern European nations (Finland, Denmark, and Sweden), none of the *Staphylococcus* strains tested could withstand exposure to this antibiotic. In certain southern European countries, such as Italy and France, resistance rates were 35–40 percent and growing rapidly, as in Germany. Penicillin resistance among pneumococci and enterococci was also on the rise.[2] As to individuals, 6.4 percent of hospital patients in France were carriers of resistant *Staphylococcus aureus*. A vicious cycle has begun: the newest antibiotics are rapidly and widely adopted in countries where resistance is common, which in turn favors the appearance of new forms of resistance. Moreover, resistant bacteria are easily transmitted from one patient to another thanks to health care settings or personnel that they have in common.

In 1994, a meeting held under the auspices of the U.S. National Foundation for Infectious Diseases resulted in the publication of recommendations with the dual purpose of restricting disease transmission and encouraging the proper usage of antibiotics. The foundation's French counterpart published an analogous document in 1996, and the battle against cross-infection between patients seems to be successful today. This success is due to the identification of patients carrying resistant bacteria and the separation of carriers who are ill from those who remain well. Equally important measures include hand-washing among caregivers, disinfection and waste disposal, and antisepsis. In the first four years of an infection control initiative at Paris Public Hospitals, strains of *Klebsiella* bacteria (a form of enteric bacteria) resistant to multiple antibiotics decreased by 72 percent in surgical units. At the same time, there was a 28 percent increase in multiply resistant *Staphylococcus* in medium- and long-term care institutions for the elderly. The success of infection control programs depends on the attention paid to the training of personnel, but it also hinges on working conditions, proper equipment in hospital rooms, and the overall organization of the hospital.

In the early 1990s, a survey of intensive care units showed that infection surveillance measures were in place in none of the units in the sample from Portugal; such measures were in place among 3 percent of French units and 77 percent of Danish units. The proportion of units with at least a washbasin in each room was 17 percent in Greece, 48 percent in France, and 74 percent in Denmark. In 44 percent of hospital units in the Netherlands, a vigilant policy of correct antibiotic usage was in place, but this was true of only 7 percent of units in France.

By the 1990s, the populations of many developed countries were increasingly obsessed by health matters and increasingly concerned by the realization that the top medical institution, the hospital, was becoming a dangerous place because of the serious infections contracted there. Outside the hospital, too, there were anxiety-provoking causes of antibiotic resistance. For example, agricultural producers used antibiotics in poultry and livestock feeds, engendering new forms of antibiotic resistance. Thus, *Pneumococcus* and *Salmonella* bacteria, among others, were gaining resistance. Antibiotics, the once-invincible weapon against bacterial infections, were losing strength over the course of the same two decades that saw the tremendous expansion of a new epidemic.

The Thunderbolt: AIDS

A new infectious disease appeared in the early 1980s, right on the heels of rising concern about the resistance of the "older" infectious diseases, which had recently yielded so readily to the astonishing power of antibiotics. Taken together, these developments threatened a return to a past that Western society had thought was gone for good. Some theories suggested that the new AIDS virus had been imported from Africa to the United States, whether directly or through the intermediary of Cuban troops in Angola; others thought that the virus came from the West to Africa with groups of scientists in the 1960s. Whatever the origins, AIDS first became manifest in the early 1980s. Mirko Grmek has provided a history of the remarkable expansion of both the illness and the medical research on the ailment.[3]

By November 1981, between 159 and 180 cases had been detected in fifteen American states, clustered particularly in neighborhoods with high concentrations of homosexual men in Los Angeles, San Francisco, and New York. Researchers from the Centers for Disease Control and Prevention in Atlanta concluded that the new syndrome was caused by an infectious agent that was spread by sexual contact. The disease did not yet have a scientific name and was referred to variously as gay pneumonia, gay cancer, or the gay plague. Then the first cases among drug addicts, including women, were reported from New York. The first European cases were young homosexuals who had visited the United States in the late 1970s. In France, between March 31 and December 29, 1982, twenty-nine cases were reported. In 1988 there were 1,384 deaths in France due to AIDS or opportunistic infections related to AIDS, or 0.3 percent of all deaths.

The impact of this epidemic went far beyond its effect on mortality in the developed countries. AIDS was, after all, the first epidemic to strike the West since cholera in the nineteenth century and the Spanish flu after World War I. Even more important was that the first victims were homosexuals and drug addicts. Old emotions resurfaced. The vilification of the marginalized, such as Haitian immigrants, homosexuals, and drug addicts, was followed by the conceptualization of the illness as a divine punishment.

In traditionalist circles in the United States, AIDS was a vindication of their resistance to all forms of sexual liberation since the 1960s. The spread of AIDS in prisons also confirmed negative stereotypes among conservatives. Fear of catching AIDS from ordinary everyday contact was acute in the general public, despite reassurances from the scientific community. Prison guards adopted odd-looking protective gear, similar to the protective outfits caricatured during nineteenth-century cholera epidemics. At the outset, the stigmatized homosexual community denied the reality of the epidemic. Then, as evidence accumulated, a change occurred around 1984. They began demanding massive public financing of research. They also worked on getting individuals to take precautions, an initiative that caused the epidemic to stabilize and ultimately to recede in the developed countries.

AIDS seemed to affect only some sectors of society, until it was discovered that blood could transmit the disease. Then the epidemic was transformed into a concern for everyone. Researchers at the Centers for Disease Control and Prevention believed that a sexually transmitted agent was to blame. It became apparent that people who tested positive on a blood test could transmit disease, and Elisa test kits were commercially available by the summer of 1985. In several countries, the government required blood donors to be screened serologically using this test. Surveys of transfusion centers in France in 1985 showed a rate of seropositivity of 0.5 to 1.0 per thousand among seemingly healthy donors. Estimates from both France and the United States showed that the number of seropositives was 50 to 100 times as great as the number of people actually sick with AIDS.

Thus arose a terrible public health problem, leading in turn to one of the great health-related scandals of the late twentieth century. Some patients were contaminated by transfusions or injections of blood products, and they accused the authorities of having reacted too slowly as scientific knowledge became available. Government decision-making processes naturally required much back-and-forth communication between scientists and policymakers, especially as they were confronted with a new epidemic, but this all seemed much too slow to those suffering the ravages of disease. The ability to identify seropositives also posed new ethical questions. How should the patient be informed of a positive test result? Should sexual partners be warned by medical personnel? Does the patient with a positive test result have the obligation to inform sexual partners? How should health care workers be protected?

On the sanitary front, there was the question of how to ensure the safety of the donated blood supply, an especially complicated problem because AIDS was not the sole concern. Hepatitis C, caused by a blood-borne virus identified in 1989, infects approximately 600,000 people today in France. All blood products for transfusions have been screened for the virus since 1990, but its prevalence in new blood donors (about 2.1 per 1,000) is higher than that of HIV (human immunodeficiency virus). The concept promulgated by Dr. Péquignot in 1953 is relevant here: successful screening for an infectious agent, HIV, is followed by a new threat, which comes to light because of a solution to an earlier crisis. In a way, this is a new definition of progress.

For the first time in history, sick people formed special interest groups and took their destiny into their own hands. They pressured government officials, scientific researchers, and the pharmaceutical industry. Some of these patients acquired a better knowledge of research advances and the latest treatments than their general doctors had. If it is true that AIDS brought back some old fears from the past, especially in the first years of the epidemic, when it seemed that infection was a death sentence, AIDS also made people rethink their attitudes about those in two main risk groups, homosexuals and drug addicts. To a significant extent, these groups escaped social disapprobation and exclusion, probably thanks to unprecedented media coverage, which was overwhelmingly favorable to AIDS research, especially in the United States.

The efforts of activist groups and public agencies to provide preventive services paid off spectacularly. The annual numbers of new cases and deaths grew in France until 1994, when they were 4,200 and 5,700, respectively. The trend reversed from 1995 onward; in 1997, there were 2,200 new cases and 1,100 deaths. Of course, various antiviral medications have increased the life expectancy among AIDS patients, but the incredible fall in the number of new cases clearly results from changes in behavior; it is entirely due to the adoption of effective preventive measures. The decline continued, so that by the year 2000 there were 1,500 cases and 550 deaths reported in France.

The slower decline in new cases as compared to the rapid decline in the death rate is probably due to the changing distribution of the illness in society. In 1992, 50 percent of seropositive test results in France were due to homosexual relations, while 21 percent were due to the use of injected drugs, 14 percent were due to heterosexual relations, and 5 percent were due to transfusions. In 2000, the numbers were 28 percent, 15 percent, 44 percent, and 0.5 percent, respectively.[4] The substantial increase in the proportion of positive tests linked with heterosexual activity demonstrates the limits of health education and prevention campaigns. In the mind of today's public, the risk of HIV is basically restricted to homosexuals and drug users. With the increasing vigilance in these two high-risk groups, heterosexuals simply do not see their own sexual activities as potentially dangerous. Another narrow view is manifest in those who say that AIDS has a historical precedent in syphilis, another sexually transmitted disease.

The AIDS epidemic was an unprecedented phenomenon in at least three ways: AIDS has a unique way of attacking the body, the development of activist organizations to fight the disease was novel, and the intensity of research efforts against the disease (after a bit of resistance) eventually reached record high levels. The progress of research was very rapid, at least with respect to the identification of the pathogenic agent and the improvement of treatments.

In the rich countries, the experience of the AIDS epidemic served to reinforce the view that the emergence of any new epidemic could only be temporary, sure to be defeated by publicity, prevention, and research advances. This view is perhaps partially correct; often an outbreak can be controlled, although it is quite another matter to eliminate a new contagious disease entirely. In the past fifteen years, the decline in deaths due to cancer and cardiovascular diseases has been much greater than the decline in deaths due to infectious diseases and respiratory conditions (although the latter two do remain relatively uncommon causes of death overall).[5] Life expectancy at birth has continued to improve, reaching 75.5 years for men and 83 years for women by 2001. Too many people still die of infectious disease, and we must not minimize that suffering. The past twenty years are characterized not only by the emergence of new infectious diseases and increasing antibiotic resistance; they are also decades during which the population's fear of infectious diseases has grown out of all proportion to reality. And overall, around the world, it has been an era of constant improvement in life expectancy at birth.

What about the Rest of the World?

The AIDS epidemic, the globalization of the economy, and expanding media have all helped to remind the West of the existence of nations in less fortunate circumstances. Europeans of the nineteenth and early twentieth centuries had been very aware of epidemics in Africa and Asia. Colonialism led to massive vaccination campaigns to protect European settlers, both civilians and the military. But the vaccination campaigns also sought to preserve the work capacity of native populations. Governor-General Carde, who was in charge of French West Africa, was alarmed by the high mortality rates there. At the opening session of the colonial Governing Council in 1927, his speech emphasized the demographic problems caused by illness.[6] Moreover, the vast colonial regions provided excellent opportunities for observations useful to military medicine as well as to Pasteurian researchers. In addition, they were vast new proving grounds for confirming the effectiveness of new treatments that had resulted from advances in bacteriology.

After the sometimes-violent process of decolonization, once-dominant military forces, including physicians, returned to their home countries. Some institutions left behind, like local Pasteur Institutes, now found themselves without logistical support. The European public focused on the enormous postwar economic expansion and the seemingly achievable eradication of infectious disease; the problems of the disadvantaged nations were ignored. But only for a while—the war in Vietnam put the third world back in the foreground, as did subsequent conflicts, which were often due to the divisive heritage of colonial rule. During the 1970 war in Biafra, 1 million people died, including many children who starved or were killed by well-known, preventable infectious diseases. Western sensibilities were reawakened to African realities by shocking media coverage that brought into the bourgeois living room the photographic and television images of children with haggard faces and distended, malnourished bellies. Médecins Sans Frontières (Doctors Without Borders), founded in 1971, has contributed to the continuing media awareness of the poor countries and their difficulties.

At present, new epidemiological threats to the rich countries seem limited, even trivial, while the day-to-day situation in the poorer nations is more like that in Europe in the nineteenth century. Infant mortality is high and there are frequent epidemics of diseases such as measles, typhoid, and diphtheria, as well as more exotic and deadlier fevers, such as Ebola. The resounding victory against infectious disease extends only as far as the boundaries of the rich nations, and investment to extend it beyond those limits, to the poorer countries, has been slow in coming. Efforts to lower the death rate from all causes and to maintain good health for all humanity are essential ingredients of the ideology of progress. They are now among the few goals on which almost everyone in the Western world can agree.

CONCLUSION

The growth of medieval cities prompted the first coordinated measures for the control of epidemics. Since then, such measures have been developed, strengthened, and systematized in the face of imminent danger, whether from plague, cholera, tuberculosis, syphilis, or AIDS. In the medieval period, local governments decided upon and administered sanitary policies. The scale and methods of controlling outbreaks changed as powerful royal governments took charge in the seventeenth century. By the nineteenth century, it was once again the municipal governments, reacting to high local mortality rates, that took the initiative to improve sanitation. Bureaus of health were city institutions established to carry out vaccination campaigns, improve public cleanliness, monitor school health, and provide health education to the workers. In Britain, Germany, Italy, and France, the work of the city health departments preceded major national health legislation.

The importance of municipal health initiatives can be illustrated by the consequences of their absence. When local vaccination campaigns against smallpox diminished during the Restoration, the results were dramatic. In the 1990s, when several republics of the former Soviet Union witnessed the effects of the complete collapse of the political system and all government health regulation, diphtheria, syphilis, and several other infectious diseases were seen once again. When infection control and health care systems lose their organizational underpinnings, such shocking reversals are the typical result. The preservation of health and the care of the body are priorities that fall by the wayside when a population is struggling to survive amid a disintegrating social fabric and severe economic uncertainty. Poverty is the enemy of good health.

Until the nineteenth century the techniques of control were focused on barriers. The goal was to prevent the epidemic from entering or spreading within the kingdom, so blockades, quarantines, and isolation hospitals were used against plague (and, later, against yellow fever and cholera). A measure of success was enjoyed as a result of these techniques, especially in Spain and France during the seventeenth and early eighteenth centuries. Then the Hippocratic view enjoyed a comeback, during the Enlightenment, and a new emphasis was placed on the role of environmental conditions in the occurrence of illness. Fresh air and water became important, and doctors and engineers worked to make them available.

The French Revolution emphasized the government's obligation to work for the good health of its citizens and stimulated the rise of the hygienist movement in France and elsewhere. The movement, dominated by physicians, was at its apogee from 1820 to 1850. For those three decades, it defined and expanded the scope of research and was recognized politically as the authoritative voice on public health matters. Now the struggle against infectious diseases was being played out in the interior of the country, and techniques were adapted accordingly: infectious disease sources had to be stamped out, particularly in the cities, by the promotion of cleanliness and by encouraging the poor to follow new and improved standards of behavior. At the same time, individual health inspections were mandated, more or less coercively depending upon the country. The degree of adherence to rules of good hygiene was monitored by home visits; immigrants were subject to health examinations; mandatory reporting and treatment of certain illnesses were established. Disinfection of lodgings was sometimes mandated, too, as was the exclusion of children from school if they had certain illnesses.

The most recent decades have been marked by expanding numbers of vaccines and by an expanding therapeutic arsenal (which consisted initially of antibiotics, followed later by antiviral medications). Taken together, these resources have seemed sufficient and appropriate to guarantee the public's health in the rich countries. Alternative techniques of disease control, such as health inspections of all arriving travelers, would be rejected as impractical; it is clear that public health is always one of many competing priorities. There are business interests to consider, as was the case during cholera epidemics. Economic interests are manifest in the recurrent concern about excessive governmental health expenditures (which has been an issue since the early nineteenth century). There are social concerns because epidemics lead to strained relations between members of different social groups. And, finally, political considerations enter into disease control measures; ever since the populations of democratic countries began to hold their governments responsible for public health, the management of outbreaks has changed dramatically.

Sanitary policies always generate resistance, and the resistance is proportional to the strictness with which the measures are applied. Protest movements, such as the anti-vaccination movement, have been very powerful in Britain and northern Europe but more limited in France—perhaps the most lax of the European countries with respect to enforcement since the late nineteenth century. The most restrictive measures—like mandatory vaccination, required reporting of infectious cases, and compulsory hospitalization—have been rejected in France in the name of civil liberties. There is a real question about the constraints that can be imposed on individuals for society as a whole to be well protected. There is also a practical matter of imposing new public health requirements little by little so that they become part of usual practice, and hostility and rejection may be minimized.

The results of public health programs in the West have been spectacular, especially in the twentieth century, although they have required considerable financial investment and massive scientific mobilization. Social welfare programs, established in Europe after World War II, offered access to medical care for the population as a whole and access to the new therapeutic discoveries provided by science. The new medicines, produced by the pharmaceutical industry, were paid for by society as a whole. The political consensus in favor of such programs is strong and deeply rooted. Since the eighteenth century, an ever-larger portion of society has sought medical attention. Between roughly 1850 and 1940, an important development took place: a substandard situation with regard to hygiene gave rise to demands for healthier conditions in the workplace and in the cities. People began to hold political figures responsible for any omissions or shortcomings in the management of public health. People began to see themselves as having rights to health, as Didier Fassin has observed.[1] Any national or local policy offering new public health measures was well received by the people, for whom any medical problem was an urgent reason to act.

On the other hand, public health measures were accompanied by a degree of social control that varied from country to country. The hygienic reformers set out to improve the habits and morality of the working class, as well as their respect for cleanliness, their lifestyle preferences, their childcare practices, and their social life. They sought to keep people from cabarets and certainly wanted to influence workers' sexual lives. Prostitutes were the object of a powerful social stigma, as were people with syphilis and even people with tuberculosis; such people were blamed for their illnesses, which were seen as the result of unhygienic behavior. And infant mortality was seen as the fault of young mothers who were poor, despite the deplorable living conditions they endured until the period between the two world wars.

On the other hand, not one country experienced a reduction in infectious disease mortality without first having a public health policy that imposed numerous constraints and involved changes in lifestyles. It's as simple as that!

Stigmatization of the lower classes extended to the populations that had been dominated by the colonial powers in the nineteenth century. The arguments offered in favor of tightening sanitary inspections at the ports of Constantinople and Alexandria demonstrate a negative attitude toward poor countries, whether Asian or African. Lack of cleanliness and periodic gathering for Moslem pilgrimages to Mecca were seen as very dangerous sources of epidemics. This attitude lasted until at least the 1950s and may be a factor even today in the West's relative indifference to and inaction in the face of medical problems in poorer countries.

Finally, our history of epidemic disease illustrates the gradual development of the physician's role in society, especially since the nineteenth century. The doctor has been in the forefront of the battle against infectious diseases and has even been the hero of the struggle when implementing and spreading the fruits of the new science of bacteriology. The luster of the doctors' role naturally faded somewhat as major epidemics receded, and doctors' status also diminished in response to the rise of diseases such as cancer and AIDS, against which physicians have less to offer. Moreover, the limitless demand in today's society for high-quality, effective medical care can only lead to disappointment when reality sets in. Lawsuits against doctors are more common than accolades.

Compared to mortality rates from infectious diseases in the 1950s, today's anxiety about Legionnaire's disease, listeria, and the like seems unwarranted. This hypersensitivity about infectious outbreaks is the result of a century and a half of the hygienists' preaching and the ongoing social pressure to comply with all sorts of everyday hygienic measures that are passed down and accentuated from generation to generation. The anxiety also stems from a general awareness (reinforced by observation of the international sanitary situation) that public health is never attained once and for all but is a permanent effort undertaken from the conviction that progress is possible. Media coverage of the health situation in poor countries reinforces this awareness. Today, the public health struggle continues on three fronts. First, the work continues to reduce the impact of infectious diseases, including "emerging" diseases, based on scientific research. Second, Western society must work to accept and integrate immigrant populations, without stigmatization, allowing them to share the goals and ideals that we have cherished for two centuries. Last, in the interest of public health, we must launch a new initiative to provide technical and economic assistance to the poorest countries.

NOTES

Chapter 1 | The Plague Era

1. Antonin Artaud, *Le Théâtre et son double* (1938; Gallimard, 1964), 33.

2. Mirko Grmek, *Les Maladies à l'aube de la civilisation occidental* (Payot, 1983).

3. Grégoire de Tours, *Historia Francorum*, 4:31.

4. Jean-Noël Biraben and Jacques Le Goff, "La Peste dans le Haut Moyen-Âge," *Annales d'histoire économique et sociale* 24 (1969): 1484–1510.

5. Peregrine Horden, "Ritual and Public Health in the Early Medieval City," in Sally Sheard and Helen Sheard, eds., *Power, Body, and City: Histories of Urban Public Health* (Ashgate, 2000), 17–40.

6. Samuel Cohn, *The Black Death Transformed: Disease and Culture in Early Renaissance Europe* (Arnold, 2001).

7. Jean-Noël Biraben, *Les Hommes et la peste en France et dans les pays européens et mediterranéens* (Mouton, 1975), 2 vols.

8. Emmanuel Le Roy Ladurie, "Un concept: L'Unification microbienne du monde (XIVe–XVIIe siècles)," *Revue d'histoire suisse* (1973), reprinted in Emmanuel Le Roy Ladurie, *Le Territoire de l'historien* (Gallimard, 1978), 2:37–97.

9. Mirko Grmek, "Préliminaires d'une étude historique des maladies," *Annales ESC* 24 (1969): 1473–1483.

10. Giorgio Cosmacini, *Soigner et réformer: Médicine et santé en Italie de la grande peste à la Première Guerre mondiale* (Payot, 1992).

11. Élisabeth Carpentier, *Une ville devant la peste: Orvieto et la peste noire de 1348* (SEVPEN, 1962).

12. Albini Giuliana, *Guerra, fame, peste: Crisi di mortalità e sistema sanitario della Lombardia tardomedioevale* (Cappelli, 1982).

13. Robert Benoît, *Vivre et mourir à Reims au Grand Siècle (1580–1720)* (Artois Presses Université, 1999).

14. Quoted in Carpentier, *Une ville devant la peste*, 33.

15. Quoted in Biraben, *Les Hommes et la peste*, 67.

Chapter 2 | Modernity

1. Carlo M. Cipolla, *Contre un ennemi invisible: Épidémies et structures sanitaires en Italie de la Renaissance au XVIIe siècle* (Balland, 1992), 109ff. (First edition was in English, 1976.)

2. Robert Benoît, *Vivre et mourir à Reims au Grand Siècle (1580–1720)* (Artois Presses Université, 1999), 67ff.

3. Jean-Noël Biraben, *Les Hommes et la peste en France et dans les pays européens et mediterranéens* (Mouton, 1975), 230ff.

4. Michel Foucault, *Histoire de la sexualité: La Volonté de savoir* (Gallimard, 1976), 1:183ff.

5. Georges Vigarello, *Histoire des pratiques de santé* (Le Seuil, 1999), 92ff.

6. Daniel Teysseire, "Un médecin dans la phase de constitution de l'hygiénisme, Louis Lépecq de La Cloture (1736–1804)," in Patrice Bourdelais, ed., *Les Hygiénistes: Enjeux, modèles, et pratiques* (Belin, 2001), 60–74.

7. Daniel Roche, *Le Siècle des Lumières en province: Académies et académiciens provinciaux* (Mouton, 1978).

8. François Lebrun, *Se soigner autrefois: Médecins, saints et sorciers aux XVIIe et XVIIIe siècles* (Messidor, 1983).

9. Françoise Loux and Richard Philippe, *Sagesses du corps: La Santé et la maladie dans les proverbes français* (Maisonneuve et Larose, 1978).

10. Olivier Faure, *Histoire sociale de la médecine (XVIIIe–XXe siècles)* (Anthropos, 1994), 28ff.

11. Alain Corbin, *Le Miasme et la jonquille: L'Odorat et l'imaginaire social, XVIIIe–XIXe siècle* (Aubier-Montaigne, 1982).

12. Hubert Charbonneau, *Tourouvre-au-Perche aux XVIIe et XVIIIe siècles: Étude de démographie historique* (INED-PUF, 1970); Alfred Perrenoud, "L'Inégalité sociale devant la mort à Genève au XVIIe siecle," *Population*, November 1975, 221–243; Hervé Le Bras and Dominique Dinet, "Mortalité des laïcs et mortalité des religieux: Les Bénédictins de Saint-Maure aux XVIIe et XVIIIe siècles," *Population*, no. 2 (1980): 347–383.

13. Yves Bayo, "Mouvement naturel de la population française de 1740 à 1829," *Population*, November 1975, 15–64; Patrice Bourdelais, *L'Âge de la vieillesse* (Odile Jacob, 1997).

14. Patrice Bourdelais, "L'Inégalité sociale face à la mort: L'Invention récente d'une réalité ancienne," in A. Leclerc, D. Fassin, H. Grandjean, M. Kaminski, and T. Lang, eds., *Les Inégalités sociales de santé* (INSERM–La Découverte, 2000), 27–39.

15. Alfred Perrenoud, "Maladies émergentes et dynamique démographique," *History and Philosophy of Life Science* 15, no. 3 (1993): 297–311.

16. Christian Hick, "Attacher les armes des mains des enfants," in Bourdelais, *Les Hygiénistes*, 41–59.

17. James Riley, "Insects and the European Mortality Decline," *American Historical Review* 3 (1986): 833–858.

18. Pierre Darmon, *La Longue Traque de la variole* (Perrin, 1986).

19. Voltaire, *Œuvres complètes* (Garnier, 1879), 22:114–115.

20. Carl Havelange, *Les Figures de la guérison (XVIIIe–XIXe siècles)* (Bibliothèque de la Faculté de Philosophie et de Lettres de l'Université de Liège, 1990), 254.

21. Peter Sköld, *The Two Faces of Smallpox* (Umeå University, 1996).

22. Norbert Elias, *La Société des individus* (Fayard, 1991). (First edition was in German, 1987.)

Chapter 3 | Cholera

1. Patrice Bourdelais and Jean-Yves Raulot, *Une peur bleue: Histoire du choléra en France, 1832–1834* (Payot, 1987).

2. Patrice Bourdelais, "La Construction de la notion de contagion: Entre médicine et société," *Communications* 66 (1998): 21–39.

3. François Delaporte, *Le Savoir de la médicine: Essai sur le choléra de 1832 à Paris* (PUF, 1990).

4. Alexandre Moreau de Jonnès, *Rapport au Conseil supérieur de santé: Le Choléra morbus pestilentiel* (Imprimerie de Cosson, 1831), 161.

5. Dr. Jachnichen, "Mémoire sur le choléra morbus qui règne en Russie, adressé de Moscou à l'Académie des sciences," *Gazette médicale*, 5 March 1831.

6. *Gazette médicale,* 2 July 1831, 11 October 1831.

7. *Gazette médicale,* 28 July 1832.

8. *Gazette médicale,* 1 October 1831.

9. Archives départementales de la Meuse, 277M5.

10. Archives départementales du Lot-et-Garonne, M5; Archives nationales, F7 9731C.

11. Patrice Bourdelais and André Dodin, *Visages du choléra* (Belin, 1987).

12. Peter Baldwin, *Contagion and the State in Europe (1830–1930)* (Cambridge University Press, 1999), chap. 2.

Chapter 4 | The "English System"

1. Christopher Hamlin, *Public Health and Social Justice in the Age of Chadwick: Britain, 1800–1854* (Cambridge University Press, 1998).

2. Louis-René Villermé, "De la mortalité dans les divers quartiers de la ville de Paris, et des causes qui la rendent très différente dans plusieurs d'entre eux, ainsi que dans les divers quartiers de beaucoup de grandes villes," *Annales d'hygiène publique et de médicine légale* 3 (1830): 294–341.

3. Peter Baldwin, *Contagion and the State in Europe (1830–1930)* (Cambridge University Press, 1999), chap. 3.

4. Milton I. Roemer, "Internationalism in Medicine and Public Health," in Dorothy Porter, ed., *The History of Public Health and the Modern State* (Rodopi, 1994), 403–423.

5. Baldwin, *Contagion and the State,* 226ff.

6. Louis-René Villermé, *Tableau de l'état physique et moral des ouvriers employés dans les manufactures de coton, de laine et de soie* (Renouard, 1840; EDI reissue 1989).

7. Claude Quétel, *Le Mal de Naples* (Seghers, 1986).

8. Alexandre Parent-Duchâtelet, *De la prostitution dans la ville de Paris* (Ballière, 1836), 2 vols.; Alain Corbin, ed., *La Prostitution à Paris au XIXe siecle* (Le Seuil, 1981).

9. Baldwin, *Contagion and the State,* chap. 5.

10. Catherine Rollet, *La Politique à l'égard de la petite enfance sous la IIIe République* (INED-PUF, 1990).

11. Pierre Budin, *Manuel pratique d'allaitement, hygiène du nourisson,* 2d ed. (Doin, 1907), 244.

12. Gaston Variot, *La Puériculture pratique* (Doin, 1913), xi.

13. Quoted in Catherine Rollet and L. Dufour, "La Goutte de lait à Fécamp, 1894–1900," *La Normandie médicale,* March 1900 (Girieud), 17.

14. Alice Reid, "Health Visitors and Child Health: Did Health Visitors Have an Impact?" *Annales de démographie historique* 1 (2001): 117–137.

15. Lara Marks, *Metropolitan Maternity: Maternal and Infant Welfare Services in Early Twentieth Century London* (Rodopi, 1996), 172.

Chapter 5 | The Sanitary Reform Movement

1. Ann La Berge, *Mission and Method: The Early Nineteenth Century French Public Health Movement* (Cambridge University Press, 1992).

2. Dora B. Weiner, *The Citizen-Patient in Revolutionary and Imperial Paris* (Johns Hopkins University Press, 1993).

3. Bernard-Pierre Lécuyer, "L'Hygiène en France avant Pasteur," in Claire Salomon-Bayet, ed., *Pasteur et la révolution pastorienne* (Payot, 1986), 65–139.

4. Florence Bourillon, "La Loi de 13 avril 1850 ou lorsque la Seconde République invente le logement insalubre," *Revue d'histoire du XIXe siècle* 1–2, nos. 20–21 (2000): 117–134.

5. Patrice Bourdelais and Jean-Yves Raulot, *Une peur bleue: Histoire du choléra en France, 1832–1854* (Payot, 1987).

6. Catherine Rollet, *La Politique à l'égard de la petite enfance sous la IIIe République* (INED-PUF, 1990), 157.

7. Stéphane Tarnier, *Recherches sur l'état puérperal* (Thèse de médicine de Paris, 1857).

8. Lion Murard and Patrick Zylberman, *L'Hygiène dans la République: La Santé publique en France ou l'utopie contrariée, 1870–1918* (Fayard, 1996), 42–49.

9. T. McKeown and R. G. Record, "Reasons for the Decline of Mortality in England and Wales during the Nineteenth Century," *Population Studies* 16 (1962): 94–122; Thomas McKeown, *The Modern Rise of Population* (London: Edward Arnold, 1976).

10. Gerry Kearns, "The Urban Penalty and the Population History of England," in A. Brandstrom and L. G. Tedebrand, eds., *Society, Health, and Population during the Demographic Transition* (Stockholm: Almqvist and Wiksell Int., 1988), 213–236; Gerry Kearns, "Le Handicap urbain et le déclin de la mortalité en Angleterre et au Pays de Galles (1851–1900)," *Annales de démographie historique* (1993): 75–105.

11. Patrice Bourdelais and Michel Demonet, "The Evolution of Mortality in an Industrial Town: Le Creusot in the Nineteenth Century," *History of the Family* 1, no. 2 (1996): 183–204.

12. Simon Szreter, "The Importance of Social Intervention in Britain's Mortality Decline c. 1850–1914: A Re-interpretation of the Role of Public Health," *Social History of Medicine* 1, no. 1 (1988): 1–37.

13. Samuel Preston and Étienne Van de Walle, "Urban French Mortality in the Nineteenth Century," *Population Studies* 32, no. 2 (1978): 275–297; Robert Millward and France Bell, "Choices for Town Councillors in Nineteenth Century Britain: Investment in Public Health and Its Impact on Mortality," in Sally Sheard and Helen Sheard, eds., *Power, Body, and City: Histories of Urban Public Health* (Ashgate: Aldershot, 2000), 143–165; Louis Cain and Elyce Rotella, "Death and Spending: Urban Mortality and Municipal Expenditure on Sanitation," *Annales de démographie historique* 1 (2001): 139–154.

14. Dr. Gibert, *Une visite au Bureau d'hygiène de Bruxelles* (Le Havre: Imprimerie F. Santallier, 1878).

15. Dr. Gibert and Lafaurie Fauvel, *Création d'un Bureau d'hygiène municipal*, proposition faite dans la séance du 11 février 1878 (Le Havre: Imprimerie Alphée Brindeau, 1878).

16. Lucie Paquy, *Santé publique et pouvoirs locaux: Le Département d'Isère et la loi du 15 février 1902*, Ph.D. diss., University of Lyon, 2001.

17. Yankel Fijalkow, *La Construction des îlots insalubres, Paris, 1850–1945* (Paris: L'Harmattan, 1998).

18. Olivier Faure, *Les Français et leur médicine au XIXe siècle* (Belin, 1993), 263.

19. Elsbeth Kalff, "Les Plaintes pour l'insalubrité du logement à Paris (1850–1950): Miroir d'hygiénisation de la vie quotidienne," in Patrice Bourdelais, ed., *Les Hygiénistes: Enjeux, modèles, et pratiques* (Belin, 2001), 118–144.

Chapter 6 | Vaccination

1. Olivier Faure, *Les Français et leur médicine au XIXe siècle* (Belin, 1993), 98ff.

2. Peter Sköld, *The Two Faces of Smallpox* (Umeå University, 1996).

3. Anne-Marie Moulin, *L'Aventure de la vaccination* (Fayard, 1996), 133–137.

4. Louis Pasteur, "Sur les maladies virulentes, et en particulier sur la maladie appelée vulgairement choléra des poules," *Comptes rendus de l'Académie des sciences* 90 (1880): 239–248.

5. Ilana Löwy, "From Guinea-pig to Man: The Development of Haffkine's Anti-cholera Vaccine," *Journal of the History of Medicine and Allied Sciences* 47 (1992): 270–309.

6. Pierre Guillaume, *Du désespoir au salut: Les Tuberculeux aux XIXe and XXe siècles* (Aubier, 1986).

7. David Barnes, "The Rise or Fall of Tuberculosis in Belle Époque France: A Reply to Allan Mitchell," *Social History of Medicine* 5, no. 2 (1992): 279–290; Arlette Mouret, "La Légende des 150,000 décès tuberculeux par an," *Annales de démographie historique*, 1996, 61–84.

8. Paul Weindling, *Health, Race, and German Politics between National Unification and Nazism, 1870–1945* (Cambridge University Press, 1989).

9. Albert Calmette, Désiré Verhaeghe, and Th. Woehrel, *Les Préventoriums ou Dispensaires de prophylaxie sociale anituberculeuse: Le Préventorium Émile-Roux de Lille* (Lille: L. Daniel, 1905).

10. Dominique Dessertine and Olivier Faure, *Combattre la tuberculose* (Lyon: PUL, 1988).

11. D. A. Hamer, *The Politics of Electoral Pressure: A Study in the History of Victorian Reform Agitations* (Hassocks, 1977).

12. David Arnold, *Imperial Medicine and Indigenous Societies* (Manchester University Press, 1988).

13. Peter Baldwin, *Contagion and the State in Europe (1830–1930)* (Cambridge University Press, 1999), 287ff.

14. Stuart M. Fraser, "Leicester and Smallpox: The Leicester Method," *Medical History* 24, no. 3 (1980): 315–332.

15. Roy Porter and Dorothy Porter, "The Politics of Prevention: Antivaccinationism and Public Health in Nineteenth Century England," *Medical History* 32 (1988): 231–252.

16. Pierre Remlinger, "Les Principaux Problèmes de la vaccination anti-rabique," *Bulletin de l'Office internationale des Épizooties* 11 (1954): 1055–1075.

17. Moulin, *L'Aventure de la vaccination*, 27–28.

Chapter 7 | The Era of Spectacular Victories

1. Pierre Guillaume, *Du Désespoir au salut: Les Tuberculeux aux XIXe and XXe siècles* (Aubier, 1986), 310ff.

2. A. Rouillon, "La Tuberculose dans le monde," *Médicine et hygiène*, April 1982; Guillaume, *Du Désespoir au salut*, 319.

3. Jacques Vallin and France Meslé, *Les Causes de décès en France de 1925 à 1978* (INED-PUF, 1988), 464.

4. Michèle Ruffat, *175 ans d'industrie pharmaceutique française* (La Découverte, 1996).

5. Sandrine Kott, *L'État social allemand* (Belin, 1995).

6. Bruno Vallat, *Histoire de la Sécurité sociale (1945–1967)* (Économica, 2001).

7. Adrien Tixier, meeting of the Comité Français de Libération Nationale, 8 February 1944, document F60 909 in the French National Archives, cited by Vallat, *Histoire de la Sécurité sociale*, 31.

8. Maurice Garden, *Histoire économique d'une grande entreprise de santé: Le Budget des hospices civils de Lyon (1800–1976)* (Presses Universitaires de Lyon, 1986), 105.

9. Henri Péquignot, *Éléments de politique et d'administration sanitaires* (ESF, 1953), 13.

Chapter 8 | The End of a Dream?

1. Serge Gottot, "Détecter, surveiller, évaluer," *Santé publique* 23 (June 1998): xxiv–xxxi.

2. R. Pallares, J. Linares, and M. Vadillo, "Resistance to Penicillin and Cephalosporin and Mortality from Severe Pneumococcal Pneumonia in Barcelona, Spain," *New England Journal of Medicine* 333 (1995): 474–480.

3. Mirko Grmek, *Histoire du SIDA* (Payot, 1989).

4. *Bulletin épidémiologique hebdomadaire*, no. 24 (2001) (Paris: Institut de Veille Sanitaire).

5. Website of the Institut national d'études démographiques (INED),

"Taux comparatifs de mortalité par grands groupes de causes de décès," produced by F. Meslé based on INSERM data.

6. Raymond Gervais, "Recensements en AOF: Genèse et signification. Des examples de la Haute-Volta coloniale," in *Annales de démographie historique* (EHESS, 1994), 339–354.

Conclusion

1. Didier Fassin, "Les Scènes locales de l'hygiènisme contemporain. La Lutte contre le saturnisme infantile: Une bio-politique à la française," in Patrice Bourdelais, ed., *Les Hygiénistes: Enjeux, modèles, et pratiques* (Belin, 2001), 447–465.

BIBLIOGRAPHY

Arnold, David. 1988. *Imperial Medicine and Indigenous Societies*. Manchester University Press.

Baldwin, Peter. 1999. *Contagion and the State in Europe, 1830–1930*. Cambridge University Press.

Barnes, David. 1995. *The Making of a Social Disease: Tuberculosis in Nineteenth-Century France*. University of California Press.

Berridge, Virginia. 1999. *Health and Society in Britain since 1939*. Cambridge University Press.

Biraben, Jean-Noël. 1975. *Les Hommes et la peste en France et dans les pays européens et mediterranéens*, 2 volumes. Mouton.

Bonah, Christian, Etienne Lepicard, and Volker Roelcke. 2003. *La Médecine expérimentale au tribunal: Implications éthiques de quelques procès médicaux du XXe siècle*. Editions Archives contemporaines.

Bourdelais, Patrice. 2001. *Les Hygiénistes: Enjeux, modèles, et pratiques*. Belin.

Bourdelais, Patrice, and Jean-Yves Raulot. 1987. *Une peur bleue: Histoire du choléra en France, 1832–1854*. Payot.

Brunton, Deborah. 2004. *Medicine Transformed: Health, Disease and Society in Europe, 1800–1930*. Manchester University Press.

Carmichael, Ann G. 1986. *Plague and the Poor in Renaissance Florence*. Cambridge University Press.

Carpenter, Kenneth J. 1986. *The History of Scurvy and Vitamin C*. Cambridge University Press.

Cipolla, Carlo M. 1992. *Contre un ennemi invisible: Épidémies et structures sanitaires en Italie de la Renaissance au XVIIe siècle*. Balland.

Cooter, Roger, and J. Pickstone, eds. 2000. *Medicine in the Twentieth Century*. Harwood.

Corbin, Alain. 1986. *The Foul and the Fragant: Odor and the French Social Imagination*. Harvard University Press.

Cosmacini, Giorgio. 1992. *Soigner et réformer: Médicine et santé en Italie de la Grande peste à la Première Guerre mondiale.* Payot.

Crosby, Alfred W. 1986. *Ecological Imperialism: The Biological Expansion of Europe, 900–1900.* Cambridge University Press.

Darmon, Pierre. 1986. *La Longue Traque de la variole.* Perrin.

Duffy, John. 1992. *The Sanitarians: A History of American Public Health.* University of Illinois Press.

Evans, Richard J. 1987. *Death in Hamburg: Society and Politics in the Cholera Years, 1830–1910.* Oxford University Press.

Fassin, Didier. 1996. *L'Espace politique de la santé.* PUF.

Faure, Olivier. 1993. *Les Français et leur médicine au XIXe siècle.* Belin.

Fee, Elizabeth, and Roy M. Acheson, eds. 1991. *A History of Education in Public Health: Health That Mocks the Doctors' Rules.* Oxford University Press.

Fee, Elizabeth, and Daniel M. Fox, eds. 1992. *AIDS: The Making of a Chronic Disease.* University of California Press.

Garrett, Laurie. 1994. *The Coming Plague.* Penguin Books.

Gélis, Jacques. 1988. *La Sage-femme ou le médicin: Une nouvelle conception de la vie.* Fayard.

Goubert, Jean-Pierre. 1989. *The Conquest of Water: The Advent of Health in the Industrial Age.* Princeton University Press.

Grmek, Mirko. 1989. *Histoire du SIDA.* Payot.

Guillaume, Pierre. 1986. *Du désespoir au salut: Les tuberculeux aux XIXe and XXe siècles.* Aubier.

Hamlin, Christopher. 1998. *Public Health and Social Justice in the Age of Chadwick: Britain, 1800–1854.* Cambridge University Press.

Hardy, Anne. 2001. *Health and Medicine in Britain since 1860.* Palgrave.

Hildreth, Martha L. 1987. *Doctors, Bureaucrats, and Public Health in France, 1888–1902.* Garland.

La Berge, Ann. 1992. *Mission and Method: The Early Nineteenth Century French Public Health Movement.* Cambridge University Press.

Lebrun, François. 1983. *Se soigner autrefois: Médecins, saints et sorciers aux XVIIe et XVIIIe siècles.* Messidor.

Léonard, Jacques. 1981. *La Médicine entre les savoirs et les pouvoirs.* Aubier.

Lewis, John. 1980. *The Politics of Motherhood: Child and Maternal Welfare, 1919–1939.* Croom Helm.

McNeill, William H. 1976. *Plagues and Peoples.* Doubleday/Anchor.

Moulin, Anne-Marie. 1996. *L'Aventure de la vaccination*. Fayard.

Phillips, Howard, and David Killingray, eds. 2003. *The Spanish Influenza Pandemic of 1918–19*. Routledge.

Porter, Dorothy. 1994. *The History of Public Health and the Modern State*. Rodopi.

Porter, Roy, and Dorothy Porter. 1988. *In Sickness and in Health: The British Experience, 1650–1850*. Blackwell.

Quetel, Claude. 1990. *History of Syphilis*. Johns Hopkins University Press.

Rodriguez-Ocana, Esteban, ed. 2002. *The Politics of the Healthy Life: An International Perspective*. European Association for the History of Medicine and Health, Sheffield.

Rollet, Catherine. 1990. *La Politique à l'égard de la petite enfance sous la IIIe République*. INED-PUF.

Rosen, George. 1993. *A History of Public Health*. Johns Hopkins University Press.

Ruffat, Michèle. 1996. *175 ans d'industrie pharmaceutique française*. La Découverte.

Sheard, Sally, and Helen Power. 2000. *Power, Body, and City: Histories of Urban Public Health*. Ashgate.

Sköld, Peter. 1996. *The Two Faces of Smallpox*. DDB, Umeö University.

Stanton, Jennifer, ed. 2002. *Innovations in Health and Medicine*. Routledge.

Starr, Paul. 1982. *The Social Transformation of American Medicine*. Basic Books.

Troesken, Werner. 2004. *Water, Race, and Disease*. MIT Press.

Vallat, Bruno. 2001. *Histoire de la Sécurité sociale (1945–1967)*. Économica.

Vigarello, Georges. 1999. *Histoire des pratiques de santé*. Le Seuil.

Weindling, Paul. 1989. *Health, Race, and German Politics between National Unification and Nazism, 1870–1945*. Cambridge University Press.

Weiner, Dora B. 1993. *The Citizen-Patient in Revolutionary and Imperial Paris*. Johns Hopkins University Press.

Woodward, John, and David Richards, eds. 1977. *Health Care and Popular Medicine in Nineteenth Century England: Essays in the Social History of Medicine*. Holmes and Meier.

INDEX